Patient-Centered Design of Cognitive Assistive Technology for Traumatic Brain Injury Telerehabilitation

Synthesis Lectures on Assistive, Rehabilitative, and Health-Preserving Technologies

Editor
Ronald M. Baecker, *University of Toronto*

Advances in medicine allow us to live longer, despite the assaults on our bodies from war, environmental damage, and natural disasters. The result is that many of us survive for years or decades with increasing difficulties in tasks such as seeing, hearing, moving, planning, remembering, and communicating.

This series provides current state-of-the-art overviews of key topics in the burgeoning field of assistive technologies. We take a broad view of this field, giving attention not only to prosthetics that compensate for impaired capabilities, but to methods for rehabilitating or restoring function, as well as protective interventions that enable individuals to be healthy for longer periods of time throughout the lifespan. Our emphasis is in the role of information and communications technologies in prosthetics, rehabilitation, and disease prevention.

Patient-Centered Design of Cognitive Assistive Technology for Traumatic Brain Injury Telerehabilitation
Elliot Cole
2013

Zero Effort Technologies: Considerations, Challenges, and Use in Health, Wellness, and Rehabilitation
Alex Mihailidis, Jennifer Boger, Jesse Hoey, and Tizneem Jiancaro
2011

Design and the Digital Divide: Insights from 40 Years in Computer Support for Older and Disabled People
Alan F. Newell
2011

Patient-Centered Design of Cognitive Assistive Technology for Traumatic Brain Injury Telerehabilitation
Elliot Cole

ISBN: 978-3-031-00466-7   print
ISBN: 978-3-031-01594-6   ebook

DOI 10.1007/978-3-031-01594-6

A Publication in the Springer series
SYNTHESIS LECTURES ON ASSISTIVE, REHABILITATIVE, AND HEALTH-PRESERVING TECHNOLOGIES
Lecture #3
Series Editor: Ronald M. Baecker, University of Toronto

Series ISSN   2162-7258 Print   2162-7266 Electronic

Patient-Centered Design of Cognitive Assistive Technology for Traumatic Brain Injury Telerehabilitation

Elliot Cole
The Institute for Cognitive Prosthetics

SYNTHESIS LECTURES ON ASSISTIVE, REHABILITATIVE, AND HEALTH-PRESERVING TECHNOLOGIES #3

ABSTRACT

This book describes a quarter-century of computing R&D at the Institute for Cognitive Prosthetics, focusing on the needs of individuals with cognitive disabilities from brain injury. Models and methods from Human Computer Interaction (HCI) can illuminate those needs, and have expanded the role of cognitive assistive technology.

User interface (UI) design has been shown to be the most important design area for individuals with cognitive disabilities. Conventional UIs typically are inappropriate for these users. Personalizing the user interface is an important strategy in achieving appropriate software for these users. Case studies are used to illustrate personalization for HCI and clinical audiences.

Patient-Centered Design (PCD) is a design methodology that incorporates both clinical and technical factors. PCD also takes advantage of the patient's ability to redesign and refine the user interface, and to achieve a very good fit between user and system. PCD can increase an individual's cognitive functioning as a technology effect. Additionally, some individuals have shown an increase in underlying cognitive abilities. Some relevant concepts from cognitive neuroscience are introduced.

Cognitive Prosthetics Telerehabilitation is a powerful therapy modality. Essential characteristics are delivering service to patients in their own home, having the patient's priority activities be the focus of therapy, using cognitive prosthetic software which applies Patient Centered Design, and videoconferencing with a workspace shared between therapist and patient.

Cognitive Prosthetics Telerehabilitation offers a rich set of advantages for the many stakeholders involved with brain injury rehabilitation.

Supplemental website, including color illustrations:

https://sites.google.com/a/morganclaypool.com/patient-centered-design/

KEYWORDS

patient-centered design; cognitive prosthetics telerehabilitation; cognitive assistive technology; cognitive prosthetics; telerehabilitation; human computer interaction; cognitive disabilities; brain injury; brain plasticity; assistive technology for cognition; cognitive rehabilitation

Dedication

This book is dedicated to Gail Kemmerer, Ph.D. and Suedell Cirwithen.

Gail started as a client, and became a collaborator, a teacher, and a friend. She was our first client, and her story is presented in Chapter 3, "Adapting Computer Software to Address Cognitive Disabilities." She taught us that an individual with severe and profound cognitive deficits can have significant cognitive abilities as well. She taught us that we could rely on the client to guide the design and redesign of the user interface, leading to high-performance intuitive designs for that individual. We learned that our users could and would create functionality from our code that went beyond the intended design. She truly was a collaborator in the design of her system, and this would become the norm. Over time she made cognitive gains and with her cognitive prosthetic software she was able to perform more and more activities without caregiver support. But mainly, over the years, she was a friend.

Suedell also started as a client. Initially she was part of a study exploring the extent of software customization needed and also exploring the use of cognitive prosthetic software as a therapy tool. Her story is presented in Chapter 4, "The Primacy of the User Interface." She thrived during the study; we continued to work with her and she became able to live independently—without caregiver support—and her recovery was considered remarkable. She was involved in the development of many therapy techniques, including therapy sessions in the home. She became an active member of the Brain Injury Association, a peer counselor, a speaker at national and international conventions, and the focus of a PBS documentary. The friendship between her family and my family continues.

Contents

Preface

This book is intended for three audiences. The first is computer science students and researchers who are interested in designing applications for individuals with cognitive disabilities, especially brain injury. The second audience is clinicians who address the needs of patients with cognitive deficits from traumatic brain injury and other diffuse types of brain injury. The third audience is the neuroscience community. Designing software for people with cognitive disabilities requires the perspectives and knowledge of all three groups.

For my colleagues in computer science, the primary message is that cognitive assistive technology can be and should be used as a tool for therapists. Our work has the power to increase the cognitive productivity of individuals with cognitive disabilities, and in some cases to increase the apparent underlying cognitive abilities of individuals with cognitive disabilities. For the clinical community, our work has shown the ability of cognitive prosthetic software to be a therapy tool. Our patient base is comprised of individuals who have come from other outpatient facilities, and an unusually high percentage of them have shown rapid gains in cognitive functioning with use of the technology, as well as some apparent increases in cognitive functioning on factors directly addressed as well as factors not directly addressed. For the neuroscience community, patients using this modality can provide subjects to study, as their gains are fairly rapid and empirical in terms of specific behavioral changes. Serial studies of these individuals who make gains may well show gains using neuroscience tools.

In 2006, I was invited to give a keynote speech to the Workshop on Cognitive Technologies at the SIGCHI annual conference. At that time, Ron Baecker, the workshop co-chair estimated that the Institute for Cognitive Prosthetics had probably treated 75% or more of the individuals treated with cognitive assistive technology. When Ron organized this series of Synthesis Lectures, he asked me to write for it.

I originally conceived this book as a means of integrating the many papers published by my colleagues and me at the Institute. I wanted to cover our years of work in the field, our technological developments, our therapy techniques, and the many stories arising from our long-term commitment to explore cognitive disabilities from brain injury.

Our work has focused on diffuse cognitive disabilities from traumatic brain injury and acquired brain injury, with a brief but fascinating diversion in reading disabilities. Three decades have passed since my first exposure to traumatic brain injury patients. During that time there have been revolutions both in personal computing and in cognitive neuroscience. Some of our early methods and findings have endured, and some have been changed due to technology advances coupled with therapists developing new techniques through on-the-job experience with our tools. Our early efforts produced many anomalous results, i.e., patient gains that were not predicted by the current medical wisdom. As time went on, advances in cognitive neuroscience could provide explanations for some of our anomalies; but our focus was on designing technology for patients and their therapists, rather than trying to engage neuroscience investigators in our efforts.

In the end, though, in writing this book, I found myself exploring new ground. The Patient-Centered Design concept presented in the 2006 Workshop was substantially refined in a paper presented at HCII in 2011, and further refined by comments and suggestions from Ron Baecker and by the reviewers. The concept of Cognitive Prosthetics Telerehabilitation (CPT) is more than an extension of our earlier model, Computer-Based Cognitive Prosthetics. In developing the CPT concept for this book, it is presented as a generic model that can be adapted to differences in national healthcare systems as well as in differences in rehabilitation organizations' structure and culture. Some of the extensions used in the Institute's implementation are described, such as the variable length therapy session. Those extensions have significance in performance measures of different parts of the system. Should other organizations adopt a version of CPT, they may well develop extensions to the CPT model that have different implications for performance of various system components. One cannot read the case studies without recognizing that patients became active and engaged in their rehabilitation therapy. Nonetheless, for years that has been tacitly but not explicitly recognized. The CPT model is structured to promote patient activity and engagement. From that came the chapter of the Active User and the Engaged User.

Some of the patient gains that have been produced by treatment with our interventions we have described as anomalies, unexplainable by what was the current clinical knowledge among cognitive rehabilitation therapists. Our mission has been apply R&D in Human Computer Interaction to extend the envelope of tools and techniques that result in observable patient behaviors. The new tools of neuroscience may be able to explain the anomalies.

Elliot Cole
Bala Cynwyd, PA
February 18, 2013

Acknowledgments

Writing about a quarter century endeavor is a complex project for an author. There are many people who have made substantial contributions to this project. Ron Baecker, the series editor, has reviewed a number of versions of a number of chapters, and his guidance has been invaluable. Diane Cerra at Morgan and Claypool has help shepherd the book through, as well as Deborah Gabriel. I have benefitted from the comments of the book's reviewers, Norman Alm, Alex Gillespie, Clayton Lewis, and Somyan Sobounov. Others who have contributed to this effort are Charline Lake, Michael Wu, and Michael Massimi, Marc Fiume, Uzma Kahn, and Paul Oramasionwu.

Special thanks goes to Parto Dehdashti, Anita Lichtenberg, Kate Wu, and Sonya Wilt for their contributions to the Institute for Cognitive Prosthetics. Others contributing to systems design are Steve Zhou, Wiley Zhu, Dan Moskowitz, Irene Glickman and Bez Thomas. Clinical contributors are Linda Petti, Marlene Angert, Richard Mahr, Michelle Zigmann, Mariann Holiday, and Candice Gustafson. Valerie Yonker focused on research. Important support was also provided by Suedell Cirwithen and Daly-Anne Leonard.

Contributors to research activity include Stanley Cole, Steven Berk, Andy Saykin, Gwen Sprehn, Peter Phillips, Nan Abbot Requel and Ruven Gur, Michael Merzenich, Marilyn Tremaine, Richard Zorowitz, Kathleen Kortte, Arnold Eiser, Sanford Schwartz, Wayne Zachary, Susan Dray, Gary Strong, Michael M. Matthews, Marilyn Spivack, George Zitney, David Jaffee, Mark Friedman, Joel Betesh, Sandra Salmans, Kit August, and Stephen Holland.

Thank you to our patients who have been our collaborators, welcoming us into their homes and lives, and who have been an inspiration to us and to those around them.

My love and thanks also go to my children, Ethan and Shoshe, Shoshe's husband Bez Thomas, and my wife Carol who has given her love, support, patience, and insight.

List of Abbreviations

ABI . . . . Acquired Brain Injury – stroke, Parkinson's, Alzheimer's, etc.

ADL . . . . Activities of Daily Living – basic activities including eating and dressing

AT . . . . Assistive Technology

CAT . . . . Cognitive Assistive Technology

CDC . . . . U.S. Centers for Disease Control and Prevention

CPT . . . . Cognitive Prosthetics Telerehabilitation – a modality of therapy – Chapter 6

HCI . . . . Human Computer Interaction

IADL . . . . Independent Activities of Daily Living – more complex daily activities

IASOD . . . . . . Islands of Abilities in Seas of Deficits

ICF . . . . International Classification of Functioning, Disability, and Health

IDSOA . . . . . . Islands of Deficits in Seas of Abilities

PCD . . . . Patient-Centered Design – a software design methodology – Chapter 5

PD . . . . Participatory Design

QI . . . . Quality Improvement

RESNA . . . . . . Rehabilitation Engineering Society of North America

R&D . . . . Research & Development

RCT . . . . Randomized Control Trial

SI . . . . Systematic Instruction

SIGCHI . . . . . . Special Interest Group/ Computer Human Interaction (and HCI)

SIGACCESS . . Special Interest Group/Accessible Computing

SSCS . . . . Single Subject Case Study

TBI . . . . Traumatic Brain Injury

TG . . . . Task Guidance

UCD . . . . User-Centered Design

UI . . . . User Interface

ZET . . . . Zero Effort Technology

CHAPTER 1

Introduction

This book is about the brain, how it functions when an individual has cognitive disabilities from brain injury, and how the field of human-computer interaction (HCI) can help increase cognitive functioning and abilities. The main focus is on how human-computer interaction and user interface (UI) design can help people with cognitive disabilities overcome those disabilities. Today's cognitive neuroscience tools allow us to study both the normal brain and the damaged brain and have led to new models of brain structure and function. Both brain structure and brain function are considerations in developing technology for traumatic brain injury (TBI) cognitive rehabilitation. In today's world, we use information and communication technology for cognitive support at work, at home, and in social and leisure activities. Mobile apps and other personal productivity tools allow us to have increased cognitive productivity. The Institute for Cognitive Prosthetics (the Institute) has expanded the use of these tools to the field of TBI rehabilitation, providing cognitive support through personal productivity tools for TBI patients.

1.1 ORIGINS OF THE WORK

In 1983, an academic consultancy led me to a tour of a residential traumatic brain injury rehabilitation facility. Patients had led normal and independent lives until their injuries, which caused multiple cognitive disabilities. Most were young. Some had been college students. Now they were incapable of living independently, and because TBI is not degenerative, most would live for decades.

Most of the patients were able to carry on a normal conversation for a while, and we could interact as peers. They all were clutching three-ring binders called memory logs, which are universal in TBI rehab. Individuals would stay in residential rehab for a period of many months, up to a year or more. These patients had a poor chance of being able to live independently. Most would leave rehab having made some progress but still facing lifelong cognitive disabilities. Their lives were permanently affected, as were the lives of their families.

In observing therapy sessions, including some in a computer lab, I came to believe that computing had the potential to help. But I also observed that patients were unable to use commercial off-the-shelf software, and thought it was obvious that patients would need therapists to guide them through the process of using the computer. A literature search found no one publishing or presenting in computer science, engineering, or psychology on the topic of computing for TBI activity support and brain injury rehabilitation.

1.2 THE PROBLEM

The problem has been to focus on the plight of individuals experiencing cognitive disabilities from TBI and to develop technology and techniques for reducing the degree of their disability, thereby increasing their self-sufficiency. This became the mission of the Institute for Cognitive Prosthetics.

A cognitive disability manifests as an inability to perform one or more everyday activities because of damage to the brain's cognitive apparatus. In the case of traumatic brain injury, a cognitive disability is the inability to perform an activity that could be performed with ease before the injury, such as making a sandwich, remembering how to use a contact list, or remembering a person's name. Most people who suffer a minor bump on the head completely recover. In contrast, a patient who suffers a TBI-induced cognitive disability will lead a life that is mildly to profoundly disrupted and will typically depend on a caregiver in order to perform some tasks.

The challenge for patients is that traditional brain injury cognitive rehabilitation has proven to be poorly effective with cognitive disabilities that do not resolve within many months as part of natural recovery. In addition, TBI damage is diffuse—it affects a broad range of cognitive abilities or dimensions. TBI also produces a unique combination of cognitive disabilities for each patient, complicating rehabilitation.

1.3 A SUCCESSFUL PILOT PROJECT

In 1987, when our research group began work, traditional TBI rehab was continuing to produce poor results, cognitive support computer software didn't exist, and specific cognitive impact differed by patient, but insurance support was substantial. TBI cognitive rehabilitation had characteristics that made it desirable from an HCI research perspective. Preliminary studies (Abbot et al.) suggested that computing functionality could be helpful to TBI patients, but the user interface was a major problem area. The user interface is basically the way the individual communicates with the computer and the computer communicates with the individual. Problem areas involved the design of commands, instructions, windows, and feedback to the user.

A pilot project was launched, involving intensive work with a single patient. That effort, although narrowly focused, highlighted the effectiveness of well-designed HCI software in cognitive rehabilitation. The initial work over a two-year period yielded significant technological and clinical success. A great deal was learned about the TBI cognitive disabilities domain, and we were successful in developing the technology that provided proof of concept (described in detail in Chapter 3). This was also the first research on cognitive disabilities to enter the computer science literature (Cole et al., 1988).

This pilot project led to four important findings. First, computer software could serve as a cognitive prosthesis. Second, the cognitively disabled user—even with profound cognitive deficits—had substantial abilities that could be applied to designing an effective cognitive assistive technology (CAT) solution. Third, the user herself was in the best position to redesign and refine the user interface (UI). Fourth, the user could exercise control over the prosthetic software features and employ them as software tools at her disposal rather than as task-guidance tools that would direct her step by step.

Developing technology with this approach has led to both (1) effective working solutions and (2) clinical and technology case studies with considerable richness of detail in descriptions of problems, solutions, and the accomplishments of the user.

1.4   THE APPROACH OF THE INSTITUTE FOR COGNITIVE PROSTHETICS

The success of the pilot led to the founding of the Institute for Cognitive Prosthetics (the Institute) in 1989. This book covers a three-decade journey by my colleagues and me at the Institute. Its mission has continued to focus on technology R&D for brain injury patients, to increase cognitive functioning and decrease dependence on caregivers. The Institute founded the fields of cognitive prosthetics and cognitive assistive technology in the late 1980s and early 1990s, (Cole et al., 1988; Cole and Dehdashti, 1990).

The Institute has taken a unique path compared to other R&D organizations working on cognitive assistive technology, with major successes technologically and clinically. That path has involved applying HCI tools with a focus on (1) CAT as a therapy tool, (2) substantial involvement of the brain injury patient in the software development process, (3) the patient's becoming, as a result, part of the design team charged with designing and revising the user interface, (4) patient use of cognitive prosthetic software in the home, and (5) application of the cognitive rehab therapy to the patient's actual priority activities.

Over the years, 21 clinicians and 10 computer scientists and developers have collaborated in working with about 100 patients. The treatment times have ranged from four months to three years, with most in the range of six months to a year.

1.5   THE NEED FOR COGNITIVE PROSTHETICS

The need for cognitive prosthetics continues to be based on cognitive disability from an injury, a disease, or another medical condition that forces the individual to rely on caregiver support or cognitive assistive technology. The impact of cognitive disabilities is amplified by the family members who must care for, or be on call for, the individual with disabilities. The impact involves reimbursed cost of care, unreimbursed care provided by people close to the individual, lost income by volunteer caregivers, lost income for the injured, stress on volunteer caregivers, and changes in lives for those involved with the injured party.

1.5.1   THE EPIDEMIOLOGY OF COGNITIVE DISABILITIES

There are a few medical conditions that are the most common causes of cognitive disability: TBI, stroke, autism spectrum disorders, senior dementias, and intellectual disabilities.

Traumatic Brain Injury

About 1.7 million people in the United States annually are seen in a hospital following a traumatic brain injury. About 52,000 die from their TBI. Of that number 275,000 are hospitalized, and about 50,000 need cognitive rehabilitation. About 1.3 million people suffer a concussion, and 90% of those resolve without treatment over the course of two weeks. There are an estimated 3.1 million adults and children living with lifelong disability from brain injury. (CDC.gov; BIAUSA.org). (Incidence data, the number of people first afflicted by a condition, is considered more reliable than prevalence data, the number of people who continue to be affected by disabilities. It is believed that prevalence estimates are low because individuals leave organizations that collect or report data.)

The Iraq and Afghanistan wars have seen substantial underreporting of TBI by about 2,500%; official statistics reported TBI diagnosed with wounds but not the impact of cumulative exposure to IEDs during the tour of duty (Schell and Marshall, 2009; Burnham et al., 2009).

Stroke

There are an estimated 1.1 million disabled stroke survivors in the United States. Each year, about 800,000 people have a stroke, and about 150,000 die from that stroke (American Stroke Association).

Autism Spectrum Disorders

The percentage of people in the United States with autism is estimated at 1% by the CDC (CDC. gov). Autism rates for adults are thought to be underreported because funding for autism services drops when an individual is no longer enrolled in educational services.

Dementias

About 6.8 million people in the United States have age-related dementia, and of that number, 1.8 million people are severely affected (Institute of Neurological Diseases and Stroke). Every five years after the age of 65, the number of people with Alzheimer's disease doubles. Alzheimer's accounts for about 70% of dementias.

Intellectual Disabilities

An estimated 6.5 million people have an intellectual disability. As with autism, the reported prevalence of intellectual disabilities drops off when educational services cease.

1.6   COGNITIVE DIMENSIONS AFFECTING THE DESIGN OF COGNITIVE ASSISTIVE TECHNOLOGY

Cognitive disabilities share a focus on cognitive functioning, but for many purposes they are heterogeneous. There are several dimensions that are helpful to the software designer in developing cognitive assistive technology CAT.

1.6.1   ACQUIRED VS. CONGENITAL

An issue in this dimension is the need to train the individual on why they may want to perform a particular activity. With an acquired disability, especially one acquired during adulthood, it is generally not necessary to teach the individual why s/he may want to perform an activity. The individual is likely to understand why, but that activity may not be important to him/her. In contrast, a congenital disability will have prevented the individual from knowing why s/he would want to perform the activity. At the very least, the individual with a congenital disability will not miss having performed the activity.

1.6.2   DEGENERATIVE VS. NON-DEGENERATIVE

This dimension is important for the amount of time the developer or researcher may have in which to develop the CAT and tailor it to a particular individual's needs. For a non-degenerative condition, the patient will not get worse during the time of software development, which can take as long as needed. For degenerative conditions, developers will need to consider the speed and character of the ongoing cognitive decline.

1.6.3   FOCAL VS. DIFFUSE

Some conditions are characterized by diffuse damage and broad impacts, while others are much more focused. TBI produces diffuse damage and broad cognitive deficits, but the combination of deficits tends to be unique, creating a Universe of One scenario. Software aimed at TBI needs to be able to support broad functionality. In contrast, strokes are focal injuries—they affect a blood vessel and tend to result in a narrower range of deficits.

1.6.4 PRIMARY VS. SECONDARY (COMPLICATIONS OF MEDICAL TREATMENT)

Some cognitive disabilities are side effects of other medical conditions or treatments. Chemotherapy for cancer can cause cognitive disabilities as a side effect. Heart-lung bypass machines for open-heart surgery can, as a side effect, cause cognitive deficits called post-perfusion syndrome.

1.6.5 SHORT-TERM VS. LONG-TERM

Many acquired and secondary conditions produce short-term cognitive deficits lasting hours to a few months. Fatigue and drugs can also cause short-term deficits. Recently, a group of HCI researchers held a workshop on the problem of short-term and transient cognitive disabilities (Gajos et al., 2012) and formed a working group on the topic. Long-term disabilities are traditional candidates for CAT.

1.7 RESPONDING TO THE NEED FOR COGNITIVE ASSISTIVE TECHNOLOGY

Science is social. A group of scientists decide that a problem is important, and that they are likely as a group to be able to make a contribution to an area. They are colleagues and competitors. To be the only person working in an area can sidetrack one's career. The plight of TBI patients who I met in 1983 was compelling, and it looked like I had a skill set involving human-centered computing, qualitative research field skills involving seeing and working in the world through others' eyes, a high comfort level with the medical area, and family experience in life-changing technology, and I was looking for colleagues. These were closely related fields. A University of Pennsylvania Medical School lab on neuropsychology and cognitive disabilities invited me to participate in their lab meetings. It was there where Andy Saykin and I struck up a friendship in the early days of neuroimaging. In the 1980s, Greg Vanderheiden was working to have various accessibility features initially as an add-on and then incorporated into the operating systems of a personal computer; there was a small but active group in ACM working on accessibility for individuals with vision and dexterity problems initially for programmers, and then for personal computing.

At approximately the same time, another group was also targeting the TBI population to support functional activities. Coming from the disciplines of electrical engineering/robotics and neuropsychology they successfully built and field-tested a portable workstation that a TBI survivor could use to perform work activities. It was designed so that an occupational therapist could program the device for a range of activities of her patient, in this project, cleaning bathrooms in a VA hospital (Kirsch et al., 1987). They called their device a "cognitive orthosis." This group was taking a similar approach to the Institute: a focus on cognitive disabilities from TBI, developing technology to compensate for disabilities, customizing the technology to the needs of the individual and designing with that individual, and using a therapist to work with the individual. There was the possibility of building a research community. Unfortunately, the group abandoned this line of research. We defined our work as "cognitive prosthetics."

The decade of the 1990s saw limited activity by other R&D groups on cognitive disabilities, but substantial activity from the Institute, and that work was welcomed at the ACM SIGCHI conferences. The "nearest neighbor" to our work was Augmentative and Alternative Communication, intended for people who are mute and who have dexterity disabilities limiting them to a couple of dozen keystrokes a minute, both physical disabilities. Work in this area involved the Rehabilitation Engineering Society of North America (RESNA). The National Head Injury Foundation, predecessor to the Brain Injury Association of America, welcomed my challenging job of recruiting participants for a technology session in their annual conferences in the 1990s. Those other groups generally followed the assistive technology paradigm. Technology was used to assist individuals in overcoming a disability. Like wheelchairs, these were devices that were used by the individual with a disability and perhaps by a caregiver (Cole, 1999).

Around the turn of the 21st century, several centers emerged with an interest in cognitive assistive technology. Among them was the well-endowed Coleman Center at the University of Colorado, and the University of Dundee group (Newell, 2011, this series; Arnott et al., 1999; Astell et al., 2010). A community of scientists was emerging, around SIGCHI and the newly formed SIGACCESS (which had emerged from the original SIGCAPH, computing and physical handicaps). Now, SIGACCESS has emerged as the principle international forum for computing related to disabilities, including cognitive disabilities. For over a decade, Ron Baecker, has been working to enhance the cognitive disabilities area. A workshop he organized for the SIGCHI 2006 meeting involved an international group of researchers. His work continues in his University of Toronto lab, Technologies for Aging Gracefully, and in founding this series of books on Assistive, Rehabilitative, and Health-Preserving Technologies.

Interest has grown in cognitive assistive technology, and there are a number of reviews of the field, also known as assistive technology for cognition. Among them are Arnott et al. (1999), LoPresti et al. (2004), LoPresti et al. (2008), Alm et. al. (2011), and Gillespie et al. (2012).

1.8   COGNITIVE DISABILITIES AS A UNIQUE DOMAIN

Cognitive disabilities span a wide range of medical diseases and conditions, with significant differences among them. However, the domain of cognitive disabilities includes an inability to perform at least two types of everyday activities such as creating a daily schedule, reacting to urgent items, writing a letter, or doing schoolwork, due to deficits in one or more cognitive dimensions. The specific activities affected will vary with each individual; any activity that a person can perform can be affected by cognitive disabilities.

Cognitive disabilities render an individual unable to perform subtasks that rely on damaged cognitive dimensions. Using computing systems involves a broad spectrum of cognitive dimensions and skills. Individuals with cognitive disabilities, however, are likely to be unable to use some of the needed dimensions and will be unable to successfully use conventional software.

The challenge is to be able to design software that both bridges cognitive impairments so that the individual can complete the subtask at hand, and also avoids using design components that rely on the impaired dimensions. This is all the more difficult because design involves fine granularity, but cognitive dimensions—as will be seen in the next chapter—are defined at a fairly coarse level of granularity, a problem also identified by Wobbrock et al. (2011).

1.8.1   COMPUTER SOFTWARE AND APPS PROVIDING BROAD COGNITIVE SUPPORT

Computing is ubiquitous and provides cognitive support for a broad array of people's everyday activities. It would take us all more time, and more involved procedures, to perform activities without the aid of computing.

The common thread of this software is that these applications help the nondisabled individual structure activities; the software does not externally and narrowly impose structure upon the activities. The individual can use these tools to plan, organize, and perform broad activities of business, household, and personal life.

Software for individuals with cognitive disabilities involves applications that relieve the individual from having to perform some cognitive subtasks, particularly those that s/he is not able to perform without caregiver support. And as we shall see later, it is significant that the individual can choose to organize an activity in many ways and can successfully complete the activity by using different commands and using them in different orders. When software tools of this sort are adapted to the needs of a cognitively impaired user, the result can be called a cognitive prosthesis (Alm et al., 2011).

1.8.2   WORKING ON ADDRESSING COGNITIVE DISABILITIES DURING REVOLUTIONS IN COMPUTATION AND NEUROSCIENCE

During the last thirty years, there has been a revolution in the use of personal computation tools such as hardware, software, communications, and services. And there has been a parallel revolution in cognitive neuroscience. The research front has been advancing quickly, and the field has seen a rapid increase in researchers. Tenets of brain structure and function that were pillars of medical science a decade ago have been undermined. However, the diffusion process is slower in the clinical areas, and a noted authority observed that research advances at the moment have little impact on the day-to-day activity of brain injury clinicians (Lezak et al., 2012).

The Institute's work began at a transitional period in both computing and neuroscience. Revolutions in computation occur rapidly. A college student can remember the technology environment during middle school and can see the changes in just a few short years. The change in technological capabilities has a profound impact on individuals and society within a period of just a few years. These same changes can also have a profound effect on CAT.

Advances in computing have had a major impact on what the Institute has been able to achieve thus far. The Institute's technology strategy was derived from Moore's Law that relatively low-cost computing systems would continue to gain hardware and software components that would significantly increase performance at lower cost. This has happened on a scale that few would have anticipated—witness, for instance, the current ubiquity of low-cost Internet service.

There have been constant incremental advances in computer technologies in the cognitive rehabilitation field. The Institute's R&D environment and culture enabled therapists to develop ideas for expanding therapy tools and techniques that would directly help them treat their patients. Researcher-developers could prototype designs, refining those with merit and discarding others. Technological advances, insightful therapists, and designers who could quickly produce prototypes worked synergistically to overcome the limitations of conventional cognitive rehabilitation. New therapy techniques allowed the exploitation of new technologies that made it practical for therapists to deliver intensive services to patient homes, incorporating patient activities into therapy so as to make therapy immediately relevant to everyday activities. New modalities of therapy evolved, first computer-based cognitive prosthetics that focused on cognitive prosthetic software, then cognitive prosthetics telerehabilitation (CPT). CPT delivers therapy to the patient's setting and superimposes therapy on top of activities that are patient priorities. This modality has many advantages over in-clinic patient therapy.

Technological advances continue in hardware, software, communications, cloud computing, and services, and these raise the prospects for new kinds of treatment and support.

Up until the 1990s, medical wisdom about the brain was not what it is today (Brooks, 1984). It was thought that brain cells could not regenerate and that the body made no new brain cells. The brain was thought to be hardwired, with all functions located in specific places. Speech was localized in the left side of the brain. Memory, attention, and other cognitive dimensions were localized in hierarchical arrangements. The window for recovery from TBI was seen as two years, while the window of recovery from stroke was thought to be one year.

Advances in cognitive neuroscience are discussed in Chapter 2.

1.8.3   FASCINATING USER OUTCOMES AND ANOMALIES

This book documents what we have learned since the mid-1980s in working with TBI patients, their therapists, their families, and in some cases their employers. Working with people who have cognitive disabilities from TBI reveals surprising results. An individual who is unable to remember one kind of information might, contrary to expectation, be able to remember other information that seems quite similar. The use of computing by TBI patients also yields surprising results, both in the kinds of tasks that they can't perform and in their ability to adapt features to their own needs in ways that they haven't been taught and that are not necessarily straightforward or obvious to observers. Unexpected developments like these arise frequently and provide data on the nuances of how the brain functions.

An anomaly in this context is a behavior that is unanticipated, given the model or information at hand. Therapists and others working closely with an individual will often spot an anomaly in the individual's behavior. Often anomalies will change some aspect of patient treatment. Anomalies can indicate that individuals have more capabilities than their clinicians and caregivers thought they had and that they can make cognitive gains faster than was thought possible.

What is remarkable is that the anomalies we are highlighting are made possible because of the use of computing. The chapters that follow will provide examples of such anomalies, along with other fascinating outcomes of the use of computing in cognitive rehabilitation.

1.9 SUMMARY

This book discusses the power of computation in treating cognitive disabilities and focuses especially on the importance of user interface design. This chapter has provided an introduction to and overview of the R&D program on traumatic brain injury rehabilitation that the author began in the 1980s. HCI models and methodologies play an important role in the the Institute's TBI cognitive rehabilitation. A surprising outcome is that user interface design can produce partial cures for TBI cognitive disabilities when used in a clinical context. This suggests that HCI may have a role in neuroscience areas related to brain plasticity.

The following chapters will provide descriptions of how these outcomes were produced.

1.10 ORGANIZATION OF THE BOOK

Chapter 2 presents "Some Clinical Features of the Cognitive Disabilities Domain with TBI Examples." This chapter will help the computer science student and researcher develop a base of knowledge of clinical issues. These issues include the difference between cognitive disabilities and cognitive deficits. There is a section on some of the many new cognitive neuroscience approaches to brain structure and function, particularly the primacy of complex networks and connectivity. There is a section on the problem of defining cognitive dimensions, with links to the World Health Organization typology of cognitive functions. The chapter ends with a strategy for dealing with cognitive constructs for the developer of software to support cognitive disabilities.

Chapter 3 is "Adapting Computer Science Models to Address Cognitive Disabilities." It provides a detailed description of working with an individual with profound cognitive disabilities and developing methodologies and application techniques. This individual changed for us from a stereotype into a person with substantial abilities as well as deficits. Most important was her ability to suggest UI design changes that worked. This was important because there was and is so little theory that can inform the design of a UI for an individual with cognitive deficits. Almost all other patients were able to participate in designing their UI using the techniques we developed with this particular patient.

Chapter 4 is "The Primacy of the User Interface." It describes why an appropriate UI is generally a barrier to computing for individuals with cognitive disabilities. The chapter goes on to provide theory on UI performance and then lays out case studies on the design of UIs and how small changes in the UI can have disproportionate impacts on UI performance.

Chapter 5 is "Patient-Centered Design" (PCD). Cognitive assistive technology can increase patients' apparent cognitive abilities and justifies using CAT as a therapist's tool. This characteristic makes CAT very different from other assistive technologies, which provide functional support but neither a partial cure nor a therapist's tool. Two case studies show the power of CAT as a therapist's tool.

Chapter 6 is "Cognitive Prosthetics Telerehabilitation," a new modality of cognitive rehabilitation therapy in which therapy is done at a distance by means of face-to-face video coupled with cognitive prosthetic software. The modality is presented in a generic form and is defined in its essential features. The technology allows new flexible and powerful therapy techniques, which have many advantages over therapy that the patient receives in the clinic. There is also a discussion of mobile technology.

Chapter 7 is "The Active User and the Engaged User." The modality of therapy developed in the previous chapters provides a structure for therapy to be adapted to make the patient active rather than passive and to promote patient emotional engagement, which has potential implications for the development of neuroscience processes in areas of deficit.

Chapter 8 is "Outcomes." A series of case studies lets the reader understand how this modality of therapy can help patients increase their ability to be independent. There are also examples of therapy failures and suggested explanations of those failures.

Chapter 9, "Conclusions, Factors Influencing Outcomes, Anomalies, and Opportunities," presents the conclusions drawn from the case studies, factors which seem to influence the outcomes, and anomalies found in the case studies. The remainder of the chapter discusses opportunities for research for the HCI and clinical communities.

CHAPTER 2

Some Clinical Features of the Cognitive Disabilities Domain with TBI Examples

This chapter explores the cognitive disabilities computing domain, with specific relevance to the design of cognitive assistive technology (CAT) and traumatic brain injury. Human Computer Interaction (HCI) and clinical areas have very different missions, approaches, methods, and cultures. These differences will become apparent as topics are discussed.

The topics include the relationship between cognitive disabilities and deficits, cognitive dimensions, the current dominant approach to brain structure and function, recent advances in neuroscience, a strategy for HCI in dealing with cognitive constructs and design of CAT, some facets of TBI rehabilitation, and a summary of relevance to the HCI researcher and practitioner.

2.1 ORIENTATION OF MEDICINE

Medical and clinical fields are largely oriented toward diseases, disorders, and pathology, in contrast to normal functioning. There is an orientation to fixing something that is "broken." Diagnosis is used to determine if there is a disease or condition, and if so, to estimate the degree of severity of the problem. As we shall see, in order to determine that there are cognitive deficits, it is not necessary to have a particularly good understanding of the functioning of the whole cognitive system.

2.2 COGNITIVE DEFICITS, COGNITIVE DISABILITIES, AND ACTIVITIES

The World Health Organization's approach to disability—the International Classification of Functioning, Disability, and Health (ICF)—has been broadly adopted by researchers and policy makers, and has also been used in cognitive assistive technology (cf. LoPresti et al., 2004 or 2008; Gillespie et al., 2012; Scherer, 2012).

A *disability* is the inability to perform an everyday *activity* because of one or more underlying cognitive deficits. And a cognitive *deficit* is divided into two components: (1) body function that is failing to perform, i.e., partial loss of memory function, as a result of (2) a bodily structure that is damaged, i.e., damage from a traumatic brain injury.

A cognitive *deficit* is damage to a cognitive function. There can be many deficits to a cognitive function, such as memory, and involve different aspects of it functioning, such as short-term memory. Clinicians summarize the damage to a cognitive function as mild, moderate, severe, and profound. From a cognitive perspective, an *activity* is a complex of cognitive functions. Deficits in one or more cognitive functions have an impact on everyday *activities*. Consequently, damage to one or more cognitive functions can be observed in an individual's inability to perform an activity; occasionally a deficit will apparently have no visible impact on functioning. A cognitive deficit is not directly observable. However, the inability to perform an everyday activity is easily observable, and then clinicians can begin to identify the cognitive functions that may be impacted, and from that identification, the brain components that may be damaged.

Cognitive functions can be seen as building blocks for an *activity*, and activity performance involves sequences of cognitive functions in parallel and in series. There are vast differences in the ways that individuals organize an activity, and these differences are reflected in the combinations of cognitive dimensions (even at low levels of granularity) that are used in performing an activity. This is sometimes referred to as cognitive style.

From this clinical perspective, an activity involves a sequence of cognitive functions almost always coupled with some physical functions. For example, handwriting involves long sequences of cognitive skills, coupled with sequences of fine motor skills.

An *activity* is well understood by HCI. An activity is composed of tasks and subtasks, on down to some elementary unit. It is important to note that subtask performance can be observed, and can be detailed to fine granularity. In contrast, cognitive dimensions cannot be directly observed, and rarely can be defined down to fine granularity. For example, problem-solving as a cognitive dimension is not subdivided into the domain of the problem, as would be done in software development.

The individual combines subtasks to form an activity. Some of these subtasks are tightly coupled in their performance, and some are loosely coupled. When timing of interrelated subtasks is important, these are called tightly coupled tasks; when there is considerable latitude in sequencing subtasks they are called loosely coupled. Rarely is a human activity so highly structured that altering the sequence of subtask performance will result in failure. There are situations in which an activity can't be performed because a subtask can't be performed. When a subtask can't be performed, the individual may be able to create a work-around that will circumvent the barrier, and reorganize the activity. Introducing computing to an activity always involves reorganizing activity performance.

There are often many different ways to organize and perform an activity. These involve using different combinations of cognitive skills, and in different sequences. An activity need not be organized precisely the same each time it is performed, and a deficit can cause a change in organization with little or no impact on efficiency of activity performance. In other instances the individual can perform the activity but at a slower pace than before the injury. There are situations where the slow pace of activity performance becomes burdensome to caregivers, who have many responsibilities to fulfill and many activities to perform in a limited period of time.

The point here is that activities can often be successfully performed when there are deficits, and the presence of deficits may not be sufficient to prohibit an activity from being performed. However, it may cause the individual to expend additional effort in order to successfully complete an activity. Furthermore, the extended period of time that an individual may require can burden caregivers who have multiple responsibilities and time pressures. Caregivers can feel the need to perform some or all of the subtasks, thereby taking away all or part of an activity that the individual can perform. This has been called "learned helplessness," which increases the individual's dependence on caregiver support.

Assistive technology (AT) can enable an individual to perform an activity when a deficit would otherwise prevent it. Without the AT, the individual wouldn't be able to participate in the activity, and that inability would amount to a disability.

2.3   SOME CLINICAL ASPECTS OF TBI

This section describes the kind of damage found in TBI patients, aspects of cognitive rehabilitation, and aspects of clinical testing.

2.3.1   DAMAGE FROM TRAUMATIC BRAIN INJURY

Damage from TBI is described in the clinical literature as diffuse, meaning that the damage is across a broad array of brain issue and brain functions (Silver et al., 2011). The damage to each function is variable, but the number of functions that are damaged increases the degree of disability, and also increases the difficulty of having successful rehabilitation. TBI rarely results in total damage to a brain function. Each injury is unique in terms of the functions affected and the degree to which each location is damaged. Moreover, an individual who receives a TBI may also have other damage caused by the same trauma, including broken bones, damage to other organs, and internal bleeding. In this way, a case may become quite complicated.

Recovering brain tissue is fragile and susceptible to additional injury. Each additional injury can cause an increasing amount of damage—the additional damage is progressive. This is especially a concern for concussion in sports and in war, where the danger of multiple concussions is greatest. In concussion, it is often difficult to detect an injury with neuroimaging, and also, it may be difficult to determine when the individual has recovered. Additionally, the culture of athletic competition and of the military has tended to minimize the seriousness of a concussion. Recent research has revealed the complex neurophysiological damage which concussion causes, and other research has shown the long-term cognitive damage caused by multiple concussions in war and in high school, college, and professional sports.

2.3.2   REHABILITATION

Cognitive disabilities are the common denominator of damage from TBI and are the focus of the largest effort in rehabilitation. Cognitive rehabilitation has changed little since the 1980s. With rare exceptions, cognitive rehabilitation is done as in-clinic therapy; in other words, the patient comes to the clinic. Conventional TBI rehab has continued to produce poor patient outcomes.

In-clinic therapy has the major disadvantage that a large number of people needing brain injury rehabilitation do not have access to it close by, because these specialized services are found near large hospitals and major medical centers. People living in rural areas, small and medium size cities, and in outer suburbs are often beyond the reasonable commuting range of brain injury rehab. TBI rehab will typically include occupational therapy, speech therapy, and physical therapy, with adjustment counseling integrated or separate.

There are two major approaches to TBI cognitive rehabilitation, both largely generic. Most rehab programs will use both, in what is described as an eclectic model of rehabilitation (National Research Council, 2011). The older approach is cognitive skills retraining, where "skills" refers to cognitive dimensions. The stategy is to build up each damaged dimension so that the patient can use them in performing her/his activities. For TBI, with its many damaged dimensions, this means building up each of the damaged dimensions for an activity to be performed. The approach has been criticized for TBI because gains achieved in one dimension become reduced when training is shifted to the next dimension. It is built around conceptions of individual cognitive skills. Re-training each skill consists of a series of computer-based exercises that are seen as progressively more difficult. Decades ago, these exercises were developed for paper-based systems, and have been ported to computer software. The cognitive areas addressed include attention and distractibility, short-term memory, working memory, auditory processing and memory, visual-spatial processing and memory including eye-hand coordination, problem-solving, information processing speed, reading, deductive and inductive reasoning, and executive functioning.

Functional rehabilitation focuses on everyday activities and was developed in the 1980s because of poor results from cognitive skills training (Mayer et al., 1986). This modality of therapy has several flavors. The most obvious is helping the individual with particular activities in his/her life. Examples include developing a schedule and planning for specific events. This modality of therapy is somewhat limited because the clinic therapist rarely visits the patient's home, and isn't able to see the patient working on their activities, and misses artifacts of activities. A second flavor involves teaching generic skills through an everyday activity. Baking brownies is often an example. The therapist can use this activity to help with reading comprehension, planning, organizing ingredients and utensils in space, sequencing, self-monitoring, as well as having brownies as a reward. In practice though, these skills are often still too abstract to address the individual's actual activities that are problematic. A third flavor is through activity simulation, which often uses worksheets and workbooks as a means of providing therapy. This flavor often doesn't account for an important subtask needed to perform the problematic activities in real life. Therapists restricted to a clinic are not able to access sufficiently detailed information about a patient's actual everyday activities to develop effective interventions. The value of cognitive rehabilitation can be subverted by generic exercises not specifically designed for the individual's specific disabilities, abilities, and activities. Context is important because TBI patients have difficulty applying abstractions. They are often unable to extract concrete applications from generic exercises. The more concrete the task that is a therapist's focus, the greater the therapist's ability to design a focused intervention and the smaller the risk of patient failure.

Computer software is rarely used as a therapy tool for directly supporting the patient's everyday activities. When we began our work in the 1980s, personal computers were not widely available, paper systems predominated in daily to-do lists, and in appointment books. Then, the value of computer software to support an individual's (and a work group, family, recreational group, etc.), activities was not well established. However, we are well into the second decade of the 21st century when such a large percentage of people rely on computers and smart devices for a great deal of cognitive support in the course of the day.

2.3.3 COGNITIVE TESTING

The main function of clinical testing for suspected brain injury is to determine if there is *evidence of pathology*, an estimate of its severity, and a diagnosis (Lezak et al., 2012).

Tests are fundamentally screening instruments. They cover a broad area of cognition with relatively few test items. Screening tests are sufficient for estimating whether there is evidence of pathology. However, they are not meant to provide accurate inventories of an individual's abilities and deficits—what the individual can and cannot do—with any substantial degree of even medium granularity. Evaluation also typically involves an interview with the patient to obtain a history.

Tests are action oriented. If there is evidence of pathology, then the results are used as a basis for obtaining clinical services, by documenting problems that are covered by insurance. Clinical testing is the gateway to receiving services paid for by insurance or other third-party payers, private or governmental. From that perspective, it is important that the results show some of the individual's abilities so that a case can be made that the individual will benefit from further treatment. The report ends by making recommendations regarding the need for additional services.

There are thousands of tests, scales, and indices that are available (Lezak et al.) and not inappropriate for assessing the patient with a suspected TBI. Clinicians will typically use a number of individual tests and scales and create a test battery. With rare exceptions, the evaluation will take place in a clinical setting and will not involve observation of the patient in his/her setting. The findings of these tests and scales, and the clinician's conclusions, are then written up in a report. In the case of TBI, daylong neuropsychological testing will cover a range of cognitive dimensions in a dozen or more tests, all at a coarse granular scale.

While most testing for cognitive disabilities uses a cognitive skills approach, there are also a substantial number of scales that evaluate an individual's ability to perform functional activities of daily living (ADLs) and independent activities of daily living (IADLs).

Unfortunately, this kind of clinical testing has little to offer the software designer. Clinical testing is done at a very coarse level of granularity, whereas software development involves a rather fine granularity of detail for the design of an application's supporting functionality and especially for the design of the user interface.

2.4 A GUIDE TO COGNITIVE DIMENSIONS FOR THE COMPUTER SCIENTIST

For the HCI practitioner or researcher, perhaps the most comprehensive and accessible approach to cognitive dimensions is the World Health Organization's International Classifications of Functioning, Disability, and Health (ICF). The ICF is accessible online at www.who.int/classifications/icf/en/ (World Health Organization, 2001). An advantage of the ICF is that it presents a hierarchical typology with terms and their definitions. Most of these are found in the ICF's section on body functioning. However, learning and interpersonal interactions—important dimensions in TBI—are located in the ICF's activities and participation section but are often treated as cognitive activities in the cognitive disabilities area.

The ICF is broadly adopted by researchers and policy makers. Computer scientists looking at it may be disappointed by its coarse granularity. This granularity is typical of the depth found in other clinical discussions, which typically lack the breadth of the ICF. While the ICF is far too general to use directly in developing user functional requirements, it does provide a framework that is useful in developing with finer granularity a set of functional descriptions that can be the basis for developing applications.

The following gives an overview of top-level cognitive functions in the ICF schema. The breadth of this typology segment is not commonly found elsewhere, especially not with definitions that help the non-psychologist specialist to understand the terms. Most typologies have fewer top-level functions. For the computer scientist, this list seems very incomplete.

- Global mental functions
- Specific mental functions
- Seeing and related functions
- Hearing and vestibular functions
- Additional sensory functions
- Pain
- Voice functions
- Articulation functions
- Fluency and rhythm-of-speech functions
- Purposeful sensory experiences
- Basic learning
- Applying knowledge
- General interpersonal interactions
- Particular interpersonal relationships
- Family relationships
- Intimate relationships

Figure 2.1: Overview of ICF cognitive function categories.

The following lists some of the more common cognitive dimensions used in TBI cognitive rehabilitation. It includes attention and memory functions. It also includes what are commonly called executive functions but under the label higher-level cognitive functions. This list provides the finest granularity for these cognitive dimensions; our developers found this granularity far too coarse for developing applications.

Figure 2.2 shows some selected common functions used in cognitive rehabilitation from the World Health Organization's International Classification of Functioning, Disability, and Health.

- Specific mental functions[1]
  - Attention functions
    - Sustaining attention
    - Shifting attention
    - Dividing attention
    - Sharing attention
  - Memory functions
    - Short-term memory
    - Long-term memory
    - Retrieval of memory
  - Thought functions
    - Pace of thought
    - Form of thought
    - Content of thought
    - Control of thought
  - Higher-level cognitive functions[2]
    - Abstraction
    - Organization and planning
    - Time management
    - Cognitive flexibility
    - Insight
    - Judgment
    - Problem-solving
  - Applying knowledge
    - Focusing attention
    - Thinking
    - Reading
    - Writing
    - Calculating
    - Solving problems
      - Solving simple problems
      - Solving complex problems

Figure 2.2: Some selected common funtions in cognitive rehab and their place in the ICF.

There is no widely used textbook that presents a broad typology of cognitive dimensions. Rather textbooks emphasize areas that are the loci of the kinds of problems typically addressed in clinical practice. For brain injury, these tend to be memory, attention, executive functioning, and language (see, for example, Sohlberg and Turkstra, 2011).

2.5   THE CURRENT CLINICAL APPROACH TO BRAIN STRUCTURE AND FUNCTIONING

Until the early 1990s, there was a dominant model of brain structure and function. The myriad functions, localized in specific places, i.e., the human brain was hard-wired. This model explained why individuals with damage to specific parts of the brain tended to have damage to particular cognitive functions or motor control. Figures 2.3 and 2.4 depict the areas of the brain where it was thought that specific functions were located. Within each part of the brain, functions were seen as hierarchically organized. These component cognitive functions could be plotted as a tree branching out to more and more specific functions.

This approach also held that, unlike all other kinds of cells that regenerate, brain cells were thought to be incapable of regeneration. It was also believed that there was a two-year window for natural cognitive recovery following a TBI and that, beyond that, there was no medical expectation of recovery. (However, insurance often paid for therapy well beyond the two-year boundary, even though significant progress was rarely seen.)

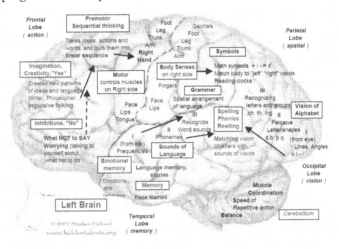

Figure 2.3: Cognitive functions and their fixed locations in the traditional model of the brain's left side.

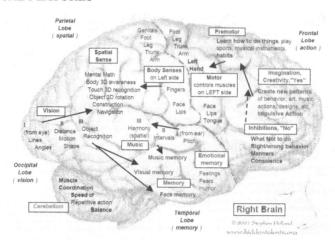

Figure 2.4: Cognitive functions and their fixed locations in the traditional model of the brain's right side.

2.5.1   COGNITIVE NEUROSCIENCE, CLINICIANS, AND CAT

My early exposure to neuroscience came from workshops sponsored by the NIH in the early and mid 1990s, whose Division of Fundamental Neuroscience was providing our funding. At one of those workshops, a leading researcher on brain plasticity suggested that the extraordinary results we had obtained with a patient could be attributed to brain plasticity; see the "Melissa" case study in Chapter 8.

Research advances in cognitive neuroscience in the last decades of the 20th Century have undermined the traditional model of brain structure and function (see Raskin, 2011). There was evidence that the speech center for a significant percentage of people could be on the right side of the brain rather than on the left, in the hard-wired paradigm. Neuroimaging studies were showing that cognitive functions were not localized to a single part of the brain, but rather involved several areas of the brain. Also, the brain is able to regenerate cells, including in an important structure related to memory, by people at least into their 70s. These advances may help explain anomalies in patient outcomes produced by the Institute's therapy techniques. These advances hold great promise for the development of improved diagnostic tools and HCI and CAT therapy tools. This section will describe the modularity of brain structure, and brain plasticity.

Modularity of the Brain's Structure

The brain is characterized by modularity and connectivity, on a massive scale and is seen as having self-organizing networks. "Virtually every perceptual or cognitive task, whether it be object recognition, memory encoding and retrieval, reading, working memory, attentional processing, motor planning or awareness, is the result of activity within large-scale and distributed brain networks" (Gazzaniga et al., 2009, page 1252). Connectivity is essential to this process, spanning the brain across lobes and hemispheres, as well as across layers of the brain. Neural circuits are seen as modular, with very specific contextual and task functions. These circuits perform activities needed for internal or external functions by combining into large distributed networks. Some network activities take place in parallel, while others take place sequentially, with timing being critical to successful functioning. The brain has a significant capacity to respond to environmental demands and in some instances can quickly form new distributed networks to adapt to new demands. Some of the networks emerge over time during the individual's development. Some networks apparently can be recruited and organized rapidly in response to environmental demands. Characteristics of objects perceived by the senses may be processed in a number of different parts of the brain, and stored in still different areas of the brain, demonstrating the modularity, specificity, and connectedness of brain circuits. There is evidence that the functional composition of these networks varies across individuals.

Brain Plasticity

Brain plasticity is evidenced by changes in aspects of complex neuronal networks responsible for a cognitive task or activity. The brain can create alternate pathways for performing a function that has been impaired. Some plasticity has no discernible impact on task function and behavior and represents an internal reorganization of the network. Neural stimulation has also been used as a means of inducing brain plasticity. Other instances of plasticity have occurred with extensive rehabilitation therapy. Recent work in cognitive neuroscience expands the scope of the concept of brain plasticity. Brain plasticity is now an accepted though limited concept in clinical areas involving brain injury and spinal cord injury. Clinicians hold out hope that brain plasticity can be a mechanism for curing injuries that have failed to heal under conventional therapies.

Cognitive neuroscience now views brain plasticity as a normal part of lifelong development rather than an exceptional occurrence as a response to injury. New neural pathways can develop relatively rapidly in response to environmental demands, and in other contexts develop slowly.

The implications of brain plasticity are enormous but also limited. First, the brain can develop alternative neural pathways for reestablishing some damaged functions, both cognitive and physical. Thus, the brain can repair itself in some instances. Second, the two-year limit for recovery of function is discredited. It would appear that there is no longer a time limit for brain plasticity to allow the development of new pathways to recover a damaged function. The process by which brain plasticity can be stimulated to restore a function is a critical research area. This is still a rapidly developing field, and there is a need for caution in applying results of individual studies as well as reviews.

2.5.2 IMPLICATIONS OF COGNITIVE NEUROSCIENCE FOR COGNITIVE ASSISTIVE TECHNOLOGY

Neuroscience holds the potential to change cognitive assistive technology from a post-therapy role to a role where CAT has an important role in therapy. Advances in cognitive neuroscience hold out the ability of individuals to continue to make gains in some aspects of cognitive functioning years and perhaps decades post injury. Various brain plasticity mechanisms hold out the potential for repair of damages to some cognitive functioning. Neuroscience is still in its early stages and it continues to expand rapidly in many directions, involving many academic disciplines.

The Institute for Cognitive Prosthetics began its R&D activities when neuroscience was not part of the clinical landscape in brain injury rehabilitation. From the beginning of our work with individuals, anomalies in their behavior were observed, patterns of behavior that were unexplained by accepted medical knowledge. In time, the fruits of neuroscience and neurophysiological research might be able to help explain some of the anomalies we had observed, and others that were becoming not unusual outcomes for our patients.

HCI researchers and designers today can begin to take advantage of neuroscience advances. Some researchers have begun using neuroimaging tools in evaluating HCI designs (Solovey et al., 2012).

CAT has the ability to increase the level of function of many individuals with cognitive disabilities from brain injury, and in some cases has demonstrated an ability to increase individuals' underlying cognitive abilities. Brain plasticity has been proposed as an explanation. However, collaboration with neuroscience researchers is needed to determine the likely biological basis for our observations.

2.6 STRATEGY FOR DEALING WITH COGNITIVE CONSTRUCTS AND TBI COGNITIVE REHABILITATION

Strange as it may sound, it is not necessary for software developers to directly address cognitive dimensions in developing cognitive assistive technology. The definition of a disability rests in barriers to activity performance. Fortunately, computing technology provides considerable cognitive support to the performance of a broad spectrum of people in all walks of life. Many computing methodologies have been developed to support the performance of activities in people's lives.

At a higher level, the strategy is to develop architecture and a development environment for addressing activities' support in the cognitive disabilities brain injury domain (Carmien et al., 2008). In that domain, a very high percentage of the Institute's development effort is spent to support personalizing the application and UI for the individual, and this personalization will continue intermittently over a period of months. Ideally, this will result in tools that occupational therapists, speech therapists, and other clinicians treating the patient will be able to use to do periodic personalization as the patient improves. In our most recent generation, personalization takes minutes.

This strategy sidesteps the measurement of an individual's cognitive functioning by focusing on artifacts of the individual's cognitive functioning—activities, subtasks, and work products. Activities and subtasks can be defined and operationalized to a fine degree of granularity, making it possible to develop user requirements that fit the individual. Cognitive deficits lack the fine granularity necessary to define the cognitive skill affected across the breadth of cognitive dimensions damaged by a TBI. More importantly, the limitation in granularity means that the gains in cognitive functioning are underestimated by being evaluated in a larger pool of dimensions. Addressing the artifacts of cognitive deficits—through their expression in activity and subtask failures—provides a workable profile of cognitive damage. This is accomplished by applying failure analysis to activity performance so that the areas needing computational support can be identified.

This strategy works with the functional rehabilitation clinical approach of TBI rehabilitation and can provide a basis for collaborating with clinicians on the treatment of patients. The strategy allows the software designer to focus on a specific case and explore the individual in depth, especially the individual's cognitive failures and abilities in a variety of contexts. This provides an excellent opportunity to explore the domain so necessary for designing effective software. TBI is a single diagnosis that masks great complexity, with unique cognitive profiles across the patient population. The ability to observe or work in depth with a number of cases helps to build domain knowledge.

2.7   SUMMARY

This chapter is intended for the HCI researcher or developer who is exploring cognitive disabilities and assistive technology. Cognitive disabilities, deficits, and cognitive skills (or dimensions) have been described and differentiated and related to individual's activities. This includes the diffuse nature of traumatic brain injury and the uniqueness of each individual's injuries; and cognitive disabilities are the common denominator of damage from TBI. Cognitive rehabilitation is a specialized in-clinic service and is broadly unavailable. Both approaches of cognitive rehab—cognitive skills retraining and functional rehabilitation—generally have poor outcomes. Clinicians do not need to have a broad, inclusive, and detailed cognitive dimensions typology in order to determine presence of pathology from TBI. However, software development requires substantial detail and specificity. This can be achieved by focusing on patient's activities rather than cognitive deficits. There was a discussion of the advances in neuroscience, and the impact of those advances on the potential for cognitive assistive technology as a therapy tool. This potential includes a technology effect of increasing cognitive functioning, along with restoring damaged cognitive abilities in some cases. Human computer interaction methodologies and models have significant opportunities for expanding opportunities for cognitive assistive technology.

With recent developments in neuroscience, there is the potential for human computer inter-action and other areas of computer science to make contributions to address cognitive disabilities. However, computer science and clinical fields have very different cultures and different views of the world. This chapter attempted to shed light on the differences. Clinical areas addressing cognitive disabilities function very well with coarse granularity in defining and operationalizing cognitive dimensions and skills. Most researchers stress that cognitive skill areas are broad systems of skills and functions that are intertwined.

Also covered in this chapter were topics such as factors affecting cognitive changes and cognitive testing.

Software development is highly detail oriented, both in concepts and in application develop-ment. This chapter defined terms such as *cognitive deficits*, *cognitive disabilities*, and *cognitive skills*. The cognitive skills and dimensions area is murky. However, developers can more easily address cognitive issues by focusing on activities in need of computational support rather than abstractions, because activities are the artifacts of cognitive functioning.

CHAPTER 3

Adapting Computer Software to Address Cognitive Disabilities

Specialists see problems differently. Clinicians have taken a look at assistive technology for cognition and seen that training the user has presented significant problems. As a result, new methods of training have been developed to instruct cognitively disabled patients to use the technology (cf. Solhberg and Turkstra, 2011). In short, the technology was seen as something that was essentially static and with a subpar fit to the user. Computer scientists and software developers look at those same problems and see a myriad of ways software can be developed or modified so that it provides a good fit rather than a subpar fit with the user (cf. Mihailidis in this series 2011).

This chapter explores the software development process for cognitive prosthetic software and is intended for a broad audience. For the human-computer interaction community, it describes the beginning of a software design methodology for an individual with cognitive disabilities from traumatic brain injury. This approach can be a good starting point in working with other cognitive disability subpopulations. For clinical communities, this provides a view into the software design process, with a focus on traumatic brain injury (TBI) patients. This process is dominated by details that come together to forge a compensatory strategy that is intuitive to the patient. The design testing process is likely to provide the therapist with new insight into the individual with cognitive impairments. This software development process works best when clinicians and designers are able to work closely together and learn from each other.

This chapter traces the initial work with our first TBI patient and describes a number of lessons learned that are still relevant today. This was the first design work in computer science on an individual with cognitive disabilities. Consequently, many procedures needed to be developed virtually from scratch. The decision to work with one individual was taken in an attempt to simplify the design process and learn about the TBI cognitive disabilities domain by looking at a single individual and a single activity to be supported. The software was intended as a post-therapy crutch for everyday activities impaired by cognitive disabilities. Much of what was learned with this individual became incorporated into Patient-Centered Design (Chapter 5) after clinicians wanted to use this technology as a therapy tool, and the approach found substantial success.

Cognitive assistive technology (CAT) for TBI works at the level of the patient's functional activity. The rationale is that cognitive disabilities affect activity performance. Activities are well understood in software development and are well suited to tracking tool use and development.

In many important ways, the functional rehabilitation model of cognitive rehabilitation is similar to the model of the design of personal support applications in computing. Both focus on providing support for the performance of people's everyday *activities* rather than on cognitive dimensions. Personal productivity tools have their conceptual origin in the 1960s, as a new class of software was defined, one that would increase an individual's cognitive productivity (Englebart, 1964; Licklider, 1960). The common characteristic is that they help individuals in facets of their everyday lives, including occupation, household, interpersonal, social, and leisure activities. But in the context of cognitive disabilities, the individual has suffered a decrease in cognitive productivity. Increasing cognitive productivity means helping the individual perform some of the everyday activities that they are now unable to perform self-sufficiently.

3.1   WORKING WITH A PATIENT: AN EXAMPLE

This example is built around our first patient because so much of the Institute's process is based on that experience (Cole and Dehdashti, 1990). One set of goals was to learn about the TBI cognitive disabilities domain, relevant knowledge from a software development perspective. Another goal was to create software that the patient could use to increase her level of functioning. The team consisted of a computer scientist, a doctoral student in HCI, and a psychologist who had been working with the patient.

Gail was four years post-TBI and had memory deficits, visual-spatial processing deficits, executive functioning deficits, and left-sided neglect syndrome. She had strong language skills, strong social skills, and strong problem-solving skills. She had a graduate education and was a professional. She was unable to make a cup of instant coffee or a simple sandwich because she was unable to sequence steps (a deficit of executive functioning). She was able to perform the subtasks of each step. (This was curious because a subtask could itself be broken down into a series of smaller subtasks.) She had loved music, but it was now too difficult for her to listen to music (a problem of executive functioning). She was diagnosed as having a left-neglect. In her case, she was aware that there was a left side of the world, but she developed a headache if she needed to concentrate on the left side of her visual field. Our initial meeting lasted just a few minutes in her therapist's office, a place where she was comfortable. She quickly put out her hand to shake hands, not knowing where her hand was in space (a visual-spatial disability) but knowing that the other person would find her hand (showing good problem-solving skills). Her language and social strengths let her appear to be normal at first. I was well aware she was far from normal, however, and I was careful not to say anything that might offend her or, given her memory problems, that she might or might not remember. (When not in a visitor's presence for a few minutes, she would forget that the visitor had been there.)

Gail had already decided that the initial application would be bill paying.[1] She had been pursuing this unsuccessfully for at least two years, in the context of therapy.

A software development methodology is the process of software development. There are different methodologies for different kinds of problems and domains. It was hoped that a software design methodology would emerge from this project, and it was expected that most steps in the process would require adaptation. We used a methodology applied to the then-emerging field of end-user computing. It would serve us for a few years and then be replaced by a method described in Chapter 6. Typically, the activity that the software is to support has already been decided. The major steps are to:

- Perform a site visit to study the functioning of the present system for that activity

- Develop a list of user requirements

  - Restructure the process

  - Design a user interface

- Identify the necessary hardware and software development tools

- Design and develop a functional module

- Design and develop the user interface

- Implement and test the functionality of the user interface

- Train the user and prepare the site

- Design and develop additional functional modules

It was clear that Gail's degree of disability would help us explore the TBI cognitive disabilities domain. Interacting with her was an immediate concern, just as it was in the initial meeting. Now we would be looking for information, and it was unclear that she would be able to provide it.

Site visits are important because they quickly reveal important factors. Site visits make it easy to get users involved in describing aspects of their problem and to allow analysts to see the problem from the users' perspective. The site visit takes place on the user's turf, making him/her more comfortable, particularly important for the TBI cognitively disabled population. There are many artifacts in the home for the analysts to view and take into account as software development progresses. Most important, other artifacts are nearby for the user to fetch and describe in a level of detail that wouldn't be possible if the interview were taking place off-site.

---

[1] Gail believed that responsible people pay their bills, and if she could pay her bills, she would be a responsible person.

Cognitive rehabilitation therapists almost never see a patient's home environment unless services are delivered to the home. Therapists typically see the clinic as the best place for the patient to receive therapy.[2]

Our plan was for the analysis to begin with a visit to Gail's home. The visit would be short and would be ended abruptly if necessary. The visit would be videotaped to allow a focus on the task at hand without the distraction of having to take notes. The video would serve as a learning tool for working with Gail, so that we could study interaction techniques to evaluate their relative effectiveness.

The visit was very productive substantively, methodologically, and interpersonally. It focused on bill paying. The aide showed the process of receiving bills, highlighting key data, and placing them in a folder for payment and then the end stages of filing the bill and mailing the payment. We asked Gail about the process, from a bill's arrival to the payment's leaving. She had a check-writing application on her computer, and she walked us through the process, retrieving artifacts along the way. She had an opportunity to show what she could do on her computer, which was basically to keyboard what was dictated to her. She also explained the difficulty she was having with the application. She had confidence in her abilities to read highlighted information on a bill and to touch-type. She had substantial difficulty reading text on the monitor. Check information spanned the screen width on a single row, and was compromised by her left-neglect. She had difficulty reading the opening menu and major difficulty understanding what she should do to respond to the software's prompts. She had inadvertently altered much of the application's data because of the ease with which the application allowed the editing of data fields. There was no audit trail—to track changes made to data—or other procedure for restoring data, so inadvertently changed or deleted data was lost forever. The application wasn't used to print out checks; they had to be made out by hand. Her visual deficits often led to checks being made out incorrectly, and she would have difficulty matching up check, bill payment form, and envelope. Sometimes the merchant's address couldn't be seen in the envelope's window.

But the most interesting observation was watching her go over some transactions on her computer. Within a few minutes, she became fatigued and had a pounding, painful headache. The headache was attributed to be physiological stress from her deficits.

Failure analysis is a technique common to many computing methodologies and strives to identify what failed and why. In the context of this project, it was valuable to see the tasks the individual could successfully perform and those she could not. It would not be necessary to do a task analysis of tasks she could do, but a detailed analysis was needed of tasks and subtasks that she could not perform. It turned out that there were several sets of issues producing task failure. She

---

[2] From the 1970s through the 1990s, it was common for patients to go from acute medical care to residential TBI rehabilitation facilities with a length of stay lasting for months. This is rare today. Perhaps the bias to the clinic is residual from these days of extended inpatient rehab, or perhaps it is administratively convenient.

could touch-type but did not know what to type or where to type it, because of the screen layout. The software had a check register, but much of the data had been compromised. The checkbook had a register, but it was difficult for her to write the check information in the narrow space provided. It was also difficult to see if payments had crossed in the mail with a new bill, which might not reflect an accurate amount due. Assembling the mailing packet—the correct envelope, payment stub correctly positioned in the envelope, and check—was very problematic. The computer didn't have a regular location, and there was no work area that was uniformly organized for her use of the computer. In the end, it was not clear what role the software played or why it was being used.

Designing a new system always means restructuring the activity. Gail desired not to depend on caregivers for the essential steps in the process. A self-sufficiency model was developed for supporting the activity despite cognitive disabilities (Fgure 3.1). Currently she required caregiver support for both core tasks—which are tightly coupled between individual and caregiver—as well as secondary tasks—ones that are loosely coupled between individual and caregiver. A tightly coupled task involves a lot of back and forth action between caregiver and the cared for. A loosely coupled task involves little action between the caregiver and the cared for. The self-sufficiency model reorganizes the task structure of a target activity, in part by introducing cognitive prosthetic software so the individual can perform an activity without relying on the availability of a caregiver. There would be core subtasks that the individual performed with computation support in a sequence of tightly coupled subtasks. Secondary activities could involve caregivers and could be performed at another time. Bill paying is an activity that is well understood computing and is highly structured. For Gail, the tightly coupled tasks would involve taking a bill, preparing the payment, and placing it in the out box. The loosely coupled tasks would be taking the check from the out box to the postal drop box and filing the check duplicate receipt along with the bill details. The actual process of taking a bill and paying it would be allocated between Gail and the application and was a tightly coupled process.

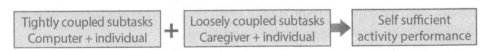

Figure 3.1: Self-sufficiency model of activity performance.

The issues of hardware and software design tools were straightforward. The IBM Personal Computer was the most powerful personal computer available with a user interface that could be modified.[3]

In selecting a design methodology, there are choices, depending in part on priorities. There are methodologies for optimizing processing efficiency, company/organizational objectives, and for focusing on "the user." A user-focused methodology could have the user be active in the design

[3] The Fat Mac had been released, but Apple's closed architecture including the UI made it inappropriate.

process, or passive. In this case, the priorities would focus on Gail, optimizing her self-sufficiency in writing checks to pay her bills. Initially, we expected Gail to be primarily passive, only responding to some of the questions posed to her. For this R&D project, we would modify user-centered design (UCD) (Gould and Lewis, 1985) to focus on an individual, Gail. She had profound and severe disabilities. She could provide information on the failures of the old system, but the accuracy of that information was suspect. Paid and family caregivers could provide information on the old system, and they could remember examples. The patient could try using the existing system and demonstrate difficulties in the use of the software. In UCD, the design team could take Gail's perspective and achieve insights that the patient couldn't be expected to have, including the causes of usage failure. UCD was consistent with having usability testing, which was anticipated to be a critical part of the software design process. The user-centered design methodology continued to be useful and could be applied with adaptations. That process would provide a guide for the computer scientists, with their process knowledge, to design to meet user requirements in light of our analysis of user performance. The computer scientists could then do the programming.

3.1.1 UI DESIGN SESSIONS

The user interface is the computing system from the perspective of the user. It determines the way the user communicates with the system and the way the system communications with the user. It also determines how the computer responds to the user's keystrokes, what is displayed on the monitor, and both audio and printed responses.

At the beginning of the project, we anticipated that work with Gail would involve iterative design, with repeated testing and design cycles. We knew that much of the design would involve hunches. Because of Gail's cognitive deficits, we decided to use on-screen mock-ups of interfaces because they were concrete and left less for the mind to fill in. Testing would first involve individual windows and then the application's UI as a whole. Gail's therapist had informed us that Gail was responsive to color and had used color in some clever compensatory strategies. Consequently, we decided to consider color as a tool in developing interfaces.

Theory suggested, and the site visit for the present system confirmed, that for an individual with cognitive disabilities from brain injury, the UI indeed could be a barrier, as it was for Gail. The UI would need to be redesigned to work around damaged cognitive skills. The challenge was that those skills were not well defined.

There were no obvious models to inform the UI design, but clinical data offered some guidance. For instance, the clinical finding of left-neglect suggested that little or no information should appear on the left side of the screen. Also, the clinically known fact that Gail had visual-spatial processing difficulty suggested that few objects should appear on each screen. She was also known to have difficulty with task sequencing—knowing or listing the order of subtasks in an activity—but was able to perform each subtask on its own. Thus the interface would need to lead her from one

subtask to another. Otherwise, clinical testing data didn't provide guidance as to color, placement of interface objects, or phrasing of commands and instructions.

Usability Test Session 1

The issues for the first session were (1) the selection of a background color (cyan), (2) the establishment of a boundary on the monitor for the left-neglect (column 42 of 80), and the determination of session length that Gail could tolerate (slightly less than half an hour).

Usability Test Session 2

The second session involved testing the interface design of a single screen. That screen would be for selecting the merchant to pay (Figure 3.2). Three kinds of objects were on the screen, all to the right of column 42: a scrolling selection box for highlighting the merchant to be paid, the ENTER key button for selecting the highlighted item, and text for instructions. The testing would apply a structured protocol to have Gail (1) point to the objects on the screen as she saw them, (2) try highlighting six of her merchants, and (3) describe how to select them. These would be timed from the video. She was able to maintain focus and remain involved in the session, a major objective of the usability test. Testing showed that, of course, some of the UI required redesign. In particular, the instruction text needed to be refined.

Figure 3.2: User interface for Usability Test 3 with Gail.

At the end of usability test session 2, Gail began commenting on some facets of the UI design, both elements that failed and those that worked. The window was redisplayed, and she gave very specific suggestions for revising the elements. The design had the ENTER key sending the command for execution, which she found difficult to understand, let alone remember. She suggested using color-coded keys to implement commands because color was so attractive and meaningful to her, especially green and red. To her, green meant continue to the next step and red meant exit. She also suggested changes in the text that she said would make it easier to understand.

This critique from Gail would become a seminal event in the development of our methodology. Our observation of traditional cognitive rehabilitation had shown it to be top-down, with therapists selecting session tasks and patients working at performing them. Because of Gail's profound and severe cognitive disabilities, our clinical collaborator had no expectation that Gail would be able to contribute to UI design, yet our user was offering suggestions to guide us from early on.

At this point, Gail had made suggestions, but their effectiveness if implemented was still in question. There were several reasons why those suggestions might not improve UI performance. The first was that Gail's cognitive deficits were so substantial that she might lack insight into her problems with using the UI; insight itself is a cognitive skill. This would mean that her suggestions lacked validity and could not usefully inform UI design. The second was that she might be able to identify problems but be unable to suggest valid solutions. Identifying problems would require insight, but suggesting solutions would require anticipating her own behavior under different conditions, a predictive accomplishment that would involve yet another set of cognitive abilities. Realistically, we thought that she might be able to anticipate some but not all of her behaviors with a revised user interface, so that additional rounds of testing and redesign would be necessary.

Working with technology, especially recent technology, will often produce complications in achieving a design; often, later versions of the technology will have removed the complications. A case in point was the decision to use color-coded function keys. Computer keyboards are laid out so that function keys are on the top row, begin on the left. Gail's left-neglect deficit meant that she would not be able to learn how to use the function keys on the left side of the keyboard.[4] There were only two or three function keys outside the left-neglect zone, and more would be needed for this application. On the other hand, the right side of the computer keyboard contained a numeric keypad laid out as if on an adding machine. We decided to try relocating the function keys to the keypad (Figure 3.3). In 1987, this was an unusual and difficult task; a few years later, it was much simpler, and today it is trivial.

---

[4] She had been a touch typist on a typewriter and had no problem using the left side of the QWERTY computer keyboard (exclusive of the function keys), same as the right. We would learn that the label left-neglect did not always apply, depending on the brain structures underlying different activities.

Figure 3.3: Keyboard with color-coded function keys assigned to the keypad Usability Test 3.

The third usability testing session presented a UI design that implemented Gail's suggestions and took her comments into account. The session protocol was changed so that there was a structured component when quantitative data could be collected and then an unstructured component when Gail could make comments and suggestions. Onscreen, there was a green box with **F6** in it (Figure 3.4), and on the keypad, there was a green key labeled **F6**. The structured part of the session showed an improvement in performance overall, and some of Gail's suggestions in the unstructured component showed us that further changes were necessary.

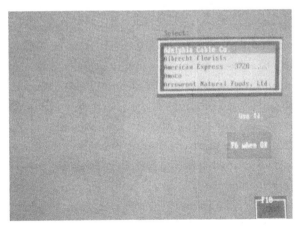

Figure 3.4: User interface revised to show a green box with **F6** in it.

Now, the unstructured component of the session was planned as the major part, during which Gail could suggest further refinements to the design to make it intuitive to her so that she could use it with a minimum cognitive load. Again she made suggestions to fix some of the problematic elements of the design (Figure 3.5). She wanted to place the green box inside a blue box, which she said would draw her attention. She suggested instruction text inside the blue box, as well as inside the green box. Figure 3.5 shows a dramatic difference in the instructions for the selection box, compared to the widget that implements the selection. The selection box contains the word Select in the upper left, and Use ↑↓ as an instruction for moving the highlight bar. In contrast, there are two phrases—Is this your choice? and F6 when OK—for what is essentially the ENTER command.

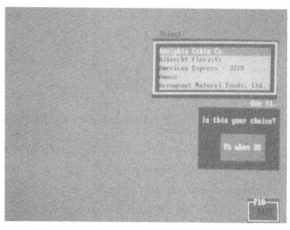

Figure 3.5: Revised screen configuration—blue box highlights green box for Gail.

The usefulness of the unstructured component of these sessions had larger and positive implications. It meant that the user with profound cognitive disabilities could conceivably not only participate in design but also play a key role in the design process. This was particularly important, as there was little theory to inform design. The ability of the cognitively disabled user to suggest solutions to design problems meant that the clinical and IT professionals would not have to invent potential solutions with little to go on. Indeed, the screen in question evolved substantially in the design surrounding the two actions in the window, from a compressed and visually muted layout (with white letters rather than the more contrasting black) to a sizable area with substantial color and contrast. It was not at all clear what design principles could be derived from this interface, but it appeared that Gail *was optimizing some factors that were important to her successful use of the interface and that indeed we were developing a methodology that could be used for this project*, although we were a long way from having a working application.

Another major UI design issue involved printing the check. During the site visit review of the present system, the impact of some of Gail's cognitive deficits was revealed. Envelopes needed to be filled so that the address on the payment coupon faced the envelope's window, with the check behind the payment coupon. She would clearly also need working space to perform the subtasks involved in stuffing an envelope and to be able to lay out the paper components. She often made errors in performing these subtasks, and there was every reason to believe that these errors would continue unless the issue of work area was resolved.

An alternative solution, however, was to create a one-page form that would bridge Gail's deficits by performing the subtasks computationally. That one-page form would be a self-seal check and mailer, and a receipt documenting the payment. The components were available: self-seal mailers, printable check forms, and magnetic ink character recognition toner for laser printers. However, the kind of mailer that we envisioned wasn't available commercially. In the mid 1980s, integrating these elements, given the software development environment for microcomputing then prevalent, would take some time. But from a design perspective, this solution was far more desirable, and the decision was made to follow this path and design the remainder of the functionality and interface accordingly. After a couple of years, however, the self-seal form was abandoned in favor of a standard check form and an envelope with a window, which she was now able to use.[5]

One of Gail's cognitive deficits was an inability to sequence subtasks in an activity. The steps in printing and correctly handling the physical check were daunting for her. Interface design could easily walk her through the steps once they were characterized. The principle issue would be identifying the steps and the cognitive chunks that she could handle, so we again relied heavily on the unstructured component of the UI testing session. Gail was both the meter for "cognitive chunk" size and the designer of instructions, which turned out to be as follows:

- Take a blank check form and load it into the printer correctly oriented, with the red dot on the left

- Confirm that the check is correctly in the printer, and send the print command

- Wait while the computer sends the text to the printer, which starts making noise while printing the document

- Retrieve the printed document

- Proofread the document for obvious printing errors

- Sign the check

- Tear off the receipt

- Fold the check and seal it

- Place a stamp on the envelope

- Place the receipt and paid bill in the to-be-filed box

- Place the envelope in the outbox to be mailed

[5] The self-seal form was being made by hand and had served its purpose. Gail was then able to separate a check and place it properly in an envelope. The check continued to show the merchant and account information, and to have the copy of the information for the filed bill.

What was surprising was the difference in chunk size across the steps. We had anticipated that the cognitive chunks would be of approximately equal size, but they were not. The first step would be the largest and most complex step by far. In that step, Gail needed to open a drawer and retrieve a blank check form, then place it in the printer's front paper bin. The step's instruction didn't even hint at the location of the blank check form. The instruction did, however, specify how to load the form (red dot on the left, using color for focus).

The UI windows contained a larger box with specific instructions, and inside that, a green box with the general instruction **F6 when done**, referring to the green **F6** key on the keypad. When Gail pressed **F6**, a tone provided feedback of step completion. When she pressed **F6** in the last step, the transaction was recorded, and the system tone generator played the first four notes of the "Halleluiah Chorus" (she had described herself as a minister's daughter).

After 17 programmer-months of effort over 10 elapsed months, the application had all the necessary functionality, and the interfaces had achieved satisfactory performance in testing. It was time to deliver the system. (Three software generations later—about a decade—it would take about two hours to prepare a patient's computer for therapy, and a few minutes to make incremental changes to patient. Thus the software easily fits within the time frame that is needed for even intensive daily therapy.)

3.1.2 INITIAL RESULTS: WHY THE CHECKS FOR $0 WERE NOT AN ERROR

After the completed application was installed, the results of the first few days were surprising and provided yet another view of the behavior of an individual with profound cognitive disabilities.

When the application was installed, Gail was able to use the application independently, paying an initial bill. She was pleased, excited, and proud at the accomplishment.

From our office, we monitored her work through the system log, intended for debugging, and through the transaction logs. She had apparently begun paying bills, but for $0, an obvious error condition. For two days we attempted to replicate the transactions but were unable to. Finally we went to her home to observe her working.

She then proudly showed us the checks. Because her previous computer hadn't worked for her, she believed it was necessary to see how well this computer could work. She realized that if she printed checks for $0, none of the balances would be affected, and she would be able to verify the merchant name, the account number, the merchant's address, and the functioning of the software. She was displaying sophisticated reasoning unexpected from an individual with some profound cognitive disabilities. Needless to say, the software testing procedure that she had come up with was not functionality that we had designed into the application but rather functionality that she had developed from the features available to her. Her caregivers assured us that this approach to testing the software came from her and not from them. This woman, with cognitive deficits so profound that five minutes after a visitor had left, she would no longer remember that an individual had visited her in her home, was able to describe the problem that she had seen and the solution she had developed.

This episode revealed an instance of the unexpected abilities of a TBI survivor with profound and severe disabilities. It also set us on notice to look for the unexpected in the future with this and with other individuals with cognitive disabilities.

This case study was presented at the 1988 SIGCHI conference, which contained a report on the first 314 days of usage. The data showed that Gail was able to use the application independently, by showing usage around the clock, including at hours when caregivers were not available to help her.

3.1.3   EPILOGUE

After about six months, Gail commented that she had been able to look in the mirror and recognize that the face in the mirror was hers as she was applying lipstick. Previously, the face was merely a target that she used to apply lipstick, but there was no emotional feeling that it was her face. Now there was. This was a major and unexpected reappearance of cognitive functioning.

We added several applications, including a word processor, an expanded contact list, and a list of places where various objects were kept in her home. With the advent of point and shoot digital cameras in the late 1990s, a multimedia album was developed.

She became able to make herself snacks and didn't need someone attending to her at every moment. She became able to use both a mouse and the Internet, behaviors and activities that were far beyond her capabilities when the computer was first introduced. She was also able to use her computer as she recovered from a heart attack and the rigors of breast cancer, including surgery, chemotherapy, and radiation.

3.2   SUMMARY

This chapter discusses aspects of software design for both the software designer and the clinician. The most important finding is proof of concept that a cognitive assistive technology application can be developed for an individual with severe and profound cognitive disabilities. Computer use is a cognitively demanding activity. It is important to note that the user was able to quickly develop novel ways of using the application to achieve needs that hadn't been anticipated by the professional design team, and the initial novel use would be considered an error condition without talking to the user. Both the initial success and the finding of novel uses became standard findings working with other TBI patients.

For the human computer interaction community, it is important that relatively conventional software development techniques could be adapted to build an application for a cognitively disabled user to perform a complex priority activity with relatively little training. This was accomplished by designing the application's user interface so that it is intuitive to that individual. Critical to this achievement was discovering that the individual with cognitive disabilities from brain injury seems to be the best meter for UI redesign and performance, with the ability to suggest, in specific terms, the redesign of UI elements. There is little else to inform UI design for cognitive disabilities from TBI, in part because of the multiplicity of factors involved in an individual's responding to a specific design. This finding has been confirmed in almost all of the Institute's patients.

For the clinician, this chapter provides a view of the very detailed design process, how it can be managed, and how the clinician can observe cognitive deficits and abilities as part of software design and use. An important step is a site visit to the patient's home to observe the current ways he performs a target activity. Failure analysis is important in developing user requirements. Where cognitive disabilities impede task performance, failure analysis will look for specific functional failures, which will need software support. Functional failures need to be understood in detail to determine the sources of the failures. Gail's case shows the level of detail that must be attended to in the process.

This chapter also develops the essence of the application-design process for cognitive prosthetic software. This involves redesigning key subtasks that are tightly coupled so that the caregiver's role is loosely coupled and the individual can perform the essence of the activity alone with the computer software.

Finally, aspects of the TBI cognitive disabilities domain were observed in the course of communicating with the patient, adapting usability sessions to the stamina of the individual, applying elements of user-centered design to software functionality, and having the individual take the lead in the design and redesign of the UI.

CHAPTER 4

The Primacy of the User Interface

This chapter describes how user interface (UI) design can produce efficient interfaces that allow individuals with cognitive disabilities to interact effectively with the world, or at least part of it, without dependence on caregivers.

The UI is the most demanding design element of computing systems from the perspective of the individual with cognitive disabilities from TBI. Using a computing system—which means using the UI—involves broad cognitive skills, and a substantial degree of precision on the part of the user. Cognitive deficits that impair performance of everyday activities also impair using computers, i.e., the user interface. UI design guidelines for the general population are often inappropriate for individuals with brain injury, who typically require a modified UI for an application to be useful at all. The UI mediates between the individual and the technology's functionality (Figure 4.1a). To the individual with cognitive disabilities, the UI can be a metaphor for communicating with the world outside (Figure 4.1b).

**For most computers, the user interface manages
the communication between the individual
and the SOFTWARE.**

Figure 4.1a: The UI mediating between the user and the software application.

**For the individual with cognitive disabilities, the
user interface manages the communication with
much of the OUTSIDE WORLD.**

Figure 4.1b: The UI metaphorically mediating between the user and the world.

Cognitive assistive technology (CAT) increases the cognitive functioning of individuals with brain injury, and UI design plays an important role in this process. This chapter uses examples from the Institute for Cognitive Prosthetics to focus on the user interface. Traumatic brain injury (TBI) is a unique domain. It is a diagnosis that covers vast combinations of cognitive dimension damage, making TBI patients an ultimately heterogeneous population in which every patient represents a Universe of One (Carmien et al., 2008). The UI needs to incorporate work-arounds for the cognitive deficits that impair any given user's ability to work with UIs designed for a general population.

Adapting UI design for individuals with cognitive disabilities presents many challenges and is a fascinating endeavor. On a technical level, a Universe-of-One domain requires a one-of-a-kind configuration. Successfully manipulating design details increases UI performance of each user. As we saw in the previous chapter, it is the user who can best inform UI design. On another level, the designer gets to observe some of the intricacies and anomalies of the injured brain in action. In particular, the observer gets to see individual strengths that often are difficult to otherwise observe. Harnessing these strengths in designing the UI is a basic strategy.

Much discussion of cognitive assistive technology focuses only on the characteristics of a specific *disability* and its impact on those people who have the disability, the same point is also made by others (cf. LoPresti et al., 2004 and Wobbrock et al., 2011). Universe-of-One domains are valuable because they focus on a specific individual, who has *abilities* in addition to disabilities. For the designer, abilities and disabilities can be viewed as opportunities to be exploited and constraints to be taken into account in developing successful cognitive work-arounds.

4.1   WHY THE TYPICAL UI IS A BARRIER TO COMPUTING USE BY INDIVIDUALS WITH BRAIN INJURY

Computer software is broadly cognitively demanding, and the field of HCI has endeavored to make it easier to use. All users have experienced some amount of difficulty in learning how to use software. It should come as no surprise, then, that individuals with cognitive disabilities—who can't independently perform some everyday activities—have even more trouble learning to use personal productivity software with conventional UIs. Conventional UIs provide particularly poor matches with TBI patients' damaged cognitive functioning.

Figure 4.2 is a heuristic model of key factors that contribute to interface performance, and key measures of performance (adapted from Card, Moran and Newell, section 12.1, 1983). This model is particularly useful in addressing cognitive disabilities. Basic usability is dependent on user characteristics (especially cognitive functioning), task characteristics, and system characteristics (which are heavily dependent on the user interface, platform characteristics, and other factors). Users who are relatively less cognitively impaired are able to tolerate limitations in interface design and other system characteristics and can still get adequate performance out of a system. These are the people who can use commercial off-the-shelf software, even with shortcomings in the UI.

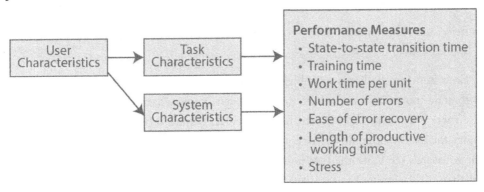

Figure 4.2: A heuristic model of UI performance for the TBI cognitive disabilities domain.

As the level of relevant user characteristics drops, as with cognitive disabilities, system characteristics, especially the UI, need to fit the individual better. Actually, the UI needs to have a much better fit with the cognitively impaired user, because the user is likely to lack some of the cognitive dimensions needed to overcome UI design flaws.

4.1.1   INTERFACE PROBLEMS OF INDIVIDUALS WITH COGNITIVE DISABILITIES FROM BRAIN INJURY

Over almost two decades, the Institute has worked intensively with individuals with cognitive disabilities from brain injury. These individuals span a broad range of impairment as well as a broad range of recovery. All initially used highly customized software, and all required substantial modification of the UI. User problems have clustered in several areas:

- Not understanding the labels for commands like Print and Save As

- Not understanding how to move from one system state to another

- Not knowing how to find specific commands in submenus

- Being overwhelmed by the UI's visual presentation and effects

- Being overwhelmed by the audio presentation and effects

- No longer understanding UI conventions such as a blinking cursor in a form field

- Not understanding instructions that are part of a form, e.g., Landscape on a print menu

- Not understanding the meaning of error messages

- Not understanding how to use the help system or the instructions in the help system

- Not understanding how to map an activity onto the application's functionality

- Being unable to process overly complex command structures that are found in bloated applications

Each item on the list is a problem of UI performance. Often the performance problem is "state-to-state transition time," where the user doesn't know what to do next and stops. This occurs because there is a mismatch between the user's current characteristics (which assumes that before the injury, the individual could use, or could easily be trained to use), and the system's characteristics, i.e., the software's UI. With the Institute's current software, it is typically faster to make changes in the UI than it is to attempt to train the individual, especially in the early stages of therapy.

Besides addressing problems, UI design can enhance an application through the use of elements like color and music that can attract or calm the user.

Remedying problems and enhancing the UI for the target user can be achieved through personalization of the UI. The most useful strategy in improving the use of software by individuals with cognitive disabilities from TBI is to simplify the UI, which in turn reduces the application's immediate functionality. Then the application's fuller functionality can be introduced slowly, as needed.

4.1.2   THE PERFORMANCE OF THE UI FOR INDIVIDUALS WITH COGNITIVE DISABILITIES FROM BRAIN INJURY

The primary role of the Institute's cognitive prosthetic software is as a platform for therapy. It doesn't serve therapy to have use of this software become a therapy goal itself with each patient. Rather, an additional therapy tool should be effective, and its use be almost seamless. The goal is, with the aid of clinicians, to increase the individual's level of functioning. Clinicians use the cognitive prosthetic software as a scaffold from which their patients can perform an activity that is part of a plan aimed at achieving a treatment goal. As a goal is achieved, a new goal is selected, and the UI is modified to provide a scaffold to help reach the new goal. At the conclusion of therapy, the prosthetic software can become a cognitive "power wheelchair," i.e., a powerful cognitive aid. At that point, it should be considered personal productivity software for a specialized market.

Whether the application is used as a platform for therapy or as a longer-term cognitive aid, an application's UI needs to deliver higher levels of performance than conventional software. Performance measures are presented in Figure 4.2. The application's functionality must be easy to learn. The application needs to be successful in helping the individual to perform a targeted activity; in performance terms, this means that work time is a relatively low number of minutes. The individual—any individual—should be expected to make a few errors, and should be able to recover from the error condition. The individual should be able to move from one system state to another without confusion. The individual should be able to use the software for a reasonable amount of time, suggesting that the UI is able to reduce the cognitive load on the individual. And finally, stress is included because it occurred with Gail, and in a good fit UI should be very small. This performance can be achieved by designing the UI to be intuitive to the individual user, i.e., by making the system's conceptual model close enough to that individual's (Lewis, 1986).

4.1.3   A STRATEGY FOR USING SOFTWARE AS A COGNITIVE PROSTHESIS

Over the years at the Institute, a strategy emerged for using software as a cognitive prosthesis. The first rule in this strategy is to collapse the time it takes the user to learn the prosthetic software. The second is to reduce functionality to what is needed for the present therapy goal. (The goals are related. Reduced functionality reduces the size and complexity of the UI and helps decrease learning time.) The third rule is to involve the user in designing and refining the UI. The fourth is to remain open to patient suggestions for added functionality so that activities can be expanded.

4.2   A STUDY IN THE IMPORTANCE OF PERSONALIZING THE UI FOR COGNITIVE PROSTHETIC APPLICATIONS

A study was designed to quantify the extent of personalization required for cognitive prosthetic applications, and the ability of those applications—coupled with therapy—to achieve a single therapy goal for each subject (Cole et al., 1994b). The study design placed an extra burden on success. In the strongest medical research design—randomized clinical trial—two methods of treatment are compared to each other to see the comparative success rate, and any complications. In this study, each research subject has already failed to achieve success with conventional therapy over an extended period of time in the target treatment goal. The question was, could this treatment approach enable a subject to succeed in a relatively short period of time, and what degree of user interface personalization was necessary to achieve success. The study was to run for three months, including training therapists. It would involve three patients, with the longest patient having about 8 weeks of therapy and the shortest four weeks of therapy, far shorter than the course of conventional therapy.

To make the study practical for a short study duration, and subject outcomes somewhat related, each therapy goal needed to be achieved using software that used a "time engine," i.e., would probably need to be related to a schedule of one kind or another.

In our study, the initial three goals were quickly met, and a decision was made to extend the number of interventions over the remaining clinical time of the study, providing more data on the UI. In all, a total of eight short-term clinical goals were achieved (see Figure 4.3).

Work with each patient began with a tour of the home by the two therapists and the lead computer scientist. The therapists had worked with the first two subjects for three and two years respectively and had never been to either's home, although one home was within walking distance of the clinic. The tour was conducted by the patient in each case and was videotaped. The visits were very revealing. The first subject, Roy, had a system of three calendars that he said served as backups for each other. The therapists were impressed by Roy's efforts in designing coping strategies and they provided some support. Roy also had a system for house keys so that he could find one before leaving his home. The second subject, Suedell, had loved to create stories to tell her children and had even prepared her own homemade books with text and drawings. On the home tour, her descriptions of storytelling suggested that she still had storytelling skills, and she also displayed some remaining drawing skills by showing us post-injury sketches. As the patients conducted their tours, therapists described both patients as functioning at a higher level than they had in the clinic.

Suedell
Initial goal:
Initiate an unsupervised activity in the home with cuing at a preset time daily; if possible, initiate an unsupervised activity spontaneously during the day
Goal 2:
Follow a brief daily schedule of activities
Goal 3:
Provide a medium of scheduled, structured writing; enhance reading activity and comprehension
Goal 4:
Increase ability to make decisions; follow increasingly complex sequence patterns

Roy
Initial goal:
Facilitate punctuality; improve ability to work with the concept of time; reduce impulsive behavior to interrupt an activity; improve attention to detail.
Goal 2:
Enable communication between patient and therapist via computer

Sarah
Initial goal:
Set priorities in the daily activities; provide memory support for activities; develop a socially appropriate compensatory strategy that could be used anytime, anywhere.
Goal 2:
Support ability to track and manage work by providing organization and structure

Figure 4.3: Patient's treatment goals for the personalization.

The software was designed with patient participation two or three days a week for about a month for Roy, with design sessions taking place in the clinic on a computer provided for the therapists. Time between sessions was needed to translate the design decisions into specifications and then to code them, using personal computing hardware and software-development tools of the early 1990s.

Two developers working part time produced 23 versions of systems for the three patients. The first system for each subject's goal was used for the initial user interface design testing, and all required redesign. These systems contained 304 interface components and 85 functional components, for a total of 389. Throughout the study, 66% of the original components had to be changed at least once to better accommodate the subject patient. More than half of the modified components needed to be changed two or more times. Participatory design was important for UI design, and a number of modifications, interestingly, either violated standard design guidelines or were counterintuitive. Almost two-thirds of the UI redesign suggestions came from either patients or clinicians, and three-fourths of the functionality changes came from patients or clinicians (Dehdashti and Cole, 1994; Cole et al., 1994b). And we noted at the time that "the impact on both patient and therapist of the opportunity to have their ideas implemented cannot be [overstated]" (Cole et al., 1994a). The therapists and patients clearly became engaged by the process.

The project's technical and clinical results provided proof of the concept that substantial and ongoing customization results in software of clinical significance for brain injury rehabilitation therapy as well as an increase in cognitive functioning. At that time, the window for cognitive recovery from TBI was judged to be two years (now modern neuroscience sees no time limit on the brain's ability for cognitive recovery from brain injury.)

It might be helpful to examine some of these interfaces and the treatment plan goals associated with them. It should be noted that the general results are applicable broadly to UI design, and are not limited to the older technologies of the early half of the 1990s when these studies were conducted.

4.2.1 SUEDELL

Suedell suffered a TBI and physical injuries five years earlier that progressed to substantial cognitive and personality changes. Her cognitive deficits affected her abstract reasoning, sequencing, cognitive flexibility, semantic memory, and activity initiation. At home she could sit for hours, inactive until someone suggested an activity. She had been outgoing and active in the community, but she became anxious, withdrawn, and fearful, with low self-confidence. She had made some gains in therapy in memory, group activities with other patients, and frustration tolerance. A teenage daughter was her primary caregiver at home.

The most productive patient in terms of goals achieved, Suedell was initially considered the most difficult to treat, because she had difficulty initiating an activity, a difficulty her therapist attributed to problems of arousal. In therapy, she would play computer games and seemed to enjoy them. The study's treatment goal was to have her play computer games at home. She responded to music and to color. Her favorite song was "Jesus Loves Me." It was decided that the computer would play "Jesus Loves Me" as a means of summoning her to the computer, centrally located in her home. She liked the idea. When asked if she would sing the song, she replied that she didn't want to hear her own voice, even though she had been a principal singer in her church's choir. Instead she just wanted to hear the song played by the computer's tone generator. During the initial usability test, she giggled with glee as the computer played the hymn.

The UI was also very colorful, which she said was stimulating to her. A large red instruction box told her to press F11 to continue to the game (Figure 4.4). The system log showed that she responded each day for the first week, and it was agreed that she had, surprisingly, achieved the intervention goal.

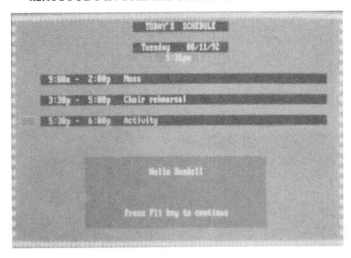

Figure 4.4: Suedell's screen at the time scheduled for Activity, when the computer played "Jesus Loves Me," calling her to her computer.

The second intervention involved a brief schedule for the day. The first intervention had used a daily schedule with a few items, which always included "Activity," indicating the time the computer would summon Suedell to the computer so that she could select her computer game. Now, with the schedule involved in the second intervention, she and her therapist would be able to plan out activities for each day, and the therapist would enter each activity into the computer using a form that was simple for the therapist but was seen as too complex for Suedell to handle on her own. But she was becoming more active and showing some initiative. When both therapists went on vacation and the lead computer scientist did the calendar remotely with the patient, she said that she believed she could enter the data on the schedule herself and proved she could.

The third intervention was a word processor (Figure 4.5). It is notable for the vibrant, even gaudy, color scheme of the basic page and the menu. This scheme was created during an interface design session with Suedell, who responded to it well. Note also the phrasing of the commands and that they are presented in a single menu. One of the commands involves deleting a document file, which is normally done at the operating system level and not at the application level. However, from Suedell's perspective deleting was part of word processor operations.

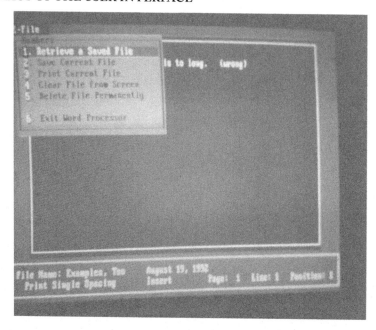

Figure 4.5: Suedell's vibrantly colored word processor.

The home visit had established that she had loved writing, especially storytelling for her six children. One of the activities was writing on a word processor. The word processor was an important application, one that she found engaging and engrossing. The study team and other therapists were impressed by her compelling stories.

This was a turning point for her. She continued to make significant progress cognitively and socially. She was able to become a mother to her daughter for the first time in years. She became active again in her church choir and became choir leader, organizing programs.

She continued to make progress and after two years could be discharged, able to live independently, but not to work. She also became active in the brain injury survivors community locally and then nationally and became a peer counselor and role model to other TBI patients. After several years, she was able to migrate from personalized prosthetic software to Microsoft Word and later to Outlook's schedule and to Internet Explorer. In short, Suedell become independent and has made a life for herself despite residual cognitive deficits.

4.2.2 ROY

Roy was a commercial artist and led an affluent lifestyle until he was mugged. An evaluation found he still had inconsistent attention and concentration, poor organization and planning, disinhibition, and poor visual memory. Significantly, he was unable to comprehend more than two weeks in advance. Prior to his enrollment in the study, he was unable to follow a routine, in part because of memory problems, or remember appointments, obligations, and his medication schedule. He also had difficulty transitioning from one activity to another.

In spite of his visual memory problems, Roy was the patient with the most visually complex interface. His first intervention was designed to get him to his activities on time, improve his ability to work with the concept of time, reduce his impulsivity in attempting to perform a second activity while still working on the first (he was afraid of forgetting), and improve his attention to detail in performing subtasks of activities. On the left panel of his UI were time slots available for activities (Figure 4.6). When an activity was highlighted in magenta, the **THINGS TO REMEMBER** box in the lower right was displayed. Roy was very successful with this application. It enabled him, during the second week, to easily be able to complete activities and move from one activity to another.

The therapist felt that an evening therapy session would be helpful in reviewing the day and planning for the next day and beyond. The therapist talked to Roy by phone, and she could see his computer via remote control software sharing his workspace. The therapist, talking with the patient, used a special editor to enter activities and things to remember.

His initial goal was declared achieved during the second week of use.

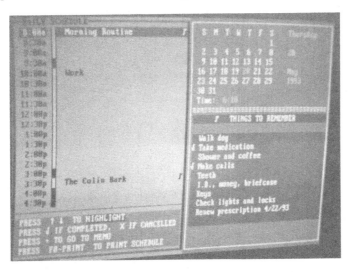

Figure 4.6: Roy's calendar. The Colin Bark (3:30 p.m.) signals his therapist that he wants to talk to her in the evening. **THINGS TO REMEMBER** in the lower right lists tasks for him to do and check off.

The nightly therapy sessions, Roy's second goal, had their origins in his desire to signal his therapist that he wished to talk to her. He problem-solved and realized that he could create a coded appointment that would signal her, and then she could respond. The appointment signal was called "The Colin Bark," after the therapist's dog. These sessions were significant for both patient and therapist. They gave the therapist a better picture of Roy's functioning in his environment and how she could make his therapy more effective. Roy became acutely aware of the importance of the computer to his daily functioning.

Perhaps the most remarkable outcome was that Roy's time horizon expanded during the first month of use from two weeks to two months. This was all the more remarkable because Roy was six years post-TBI at the time of this study. He was able to plan ahead for social, financial, and medical activities. He made substantial gains in other areas as well, including problem solving and task tracking, so that he didn't need to interrupt one task when another popped into his mind. The nightly therapy sessions helped him prioritize, think ahead, and better estimate the amount of time needed for a task.

4.2.3 SARAH

Sarah, a 27-year-old woman, was involved in a minor motor vehicle accident and suffered post-concussion syndrome. She was a graduate student in a professional program. A year-and-a-half after the accident she still had poor organization, difficulty prioritizing, and poor concentration. These deficits had an impact on her ability to track appointments, pay bills, track to-do items, and prioritize. Worrying about these, in the context of her profession, caused episodes of reactive depression, which could cause a downward spiral in her cognitive and emotional functioning and physical stamina.

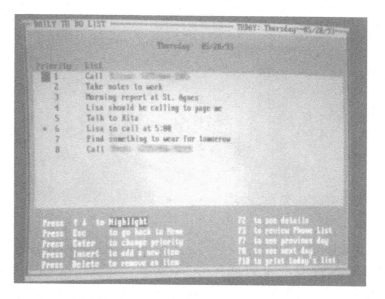

Figure 4.7: One of Sarah's errand lists. She was unable to priority the items in even a short list as a batch. However, she could reprioritize one item at a time, and this application displayed the reordered list after each item change. Note dense list of instructions at the bottom of the window.

In a pretest, Sarah generated a list of half a dozen errands she needed to do. She was then unable to prioritize any in two minutes, and the pretest was terminated. She was, however, able to quickly prioritize the list one item at a time, and this became the basis of her first goal. She took control of application design and specified how she wanted the application to function and to look (see Figure 4.7). The application's most important functionality was the one-item-at-a-time prioritization method, with the screen refreshed and re-ordered after each priority change. The priority field was implemented by using the time-field for an item, and the time-sort feature of the "time engine." Each priority was assigned an hour and minute, which was changed whenever Sarah changed the priority. Each errand item was displayed as a line, with the rank order being its position in the daily array.

The ten available commands are shown at the bottom of the window. Visually this documentation is very dense, and one wonders if it is too dense for someone with cognitive deficits, particularly in processing visual information. One suspects that the deficits that impair visually reordering lines of text are independent of the cognitive dimensions involved in processing the list of instructions.

She achieved the goal of prioritizing and completing a daily list during the first week. Over her 50-day course of the study, she used the system 45 days, with an average of three items per day.

Her prioritizing also seemed to be an anomaly. She had been unable to begin to prioritize a list of only six items, yet she could quickly prioritize items one at a time. That seemed surprising, given her intelligence. Computer technology provided a perfect solution for reranking the items after each change.

Shortly after the initial success, she asked for a case-tracking application, her second goal. Again, she designed the application's features and UI (see Figure 4.8). She used the case-tracking system for ten clients over the course of a month. A couple of her colleagues, seeing her efficiency and examining the application, requested copies for themselves. This demonstrated that it was not stygmatizing.

These applications were important to Sarah because she felt they created a floor for her cognitive functioning and her reactive depression episodes. She knew that she was able to function with these applications and that she would avoid the reactive depression episodes she had experienced in the past.

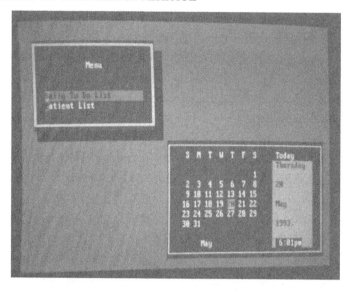

Figure 4.8: Main menu for Sarah after the second application was added.

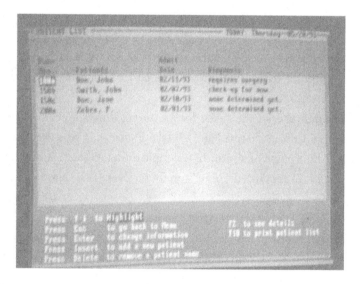

Figure 4.9: A screen from Sarah's second application.

4.2.4 SIX ADDITIONAL SUBJECTS

Six additional subjects were added to the study. For this six, the issue was whether cognitive prosthetic software could achieve the goal set for subjects who had failed to achieve the goal with conventional software, and if so, how long did it take. For the six additional subjects, no data was collected on interface or functional modules requiring modification.

The total sample was nine, and seven—78%—were able to achieve their goal within three weeks (Cole et al, 1997). These results were considered very successful because the study population was composed of people who had failed to achieve the goal after at least months of conventional cognitive rehabilitation. This provided additional support for the power of software personalized to the needs of the specific patient.

4.2.5   SUMMARY: THE IMPORTANCE OF PERSONALIZING THE UI STUDY

This study demonstrated the power of user interface design in helping brain injury patients achieve success in a therapy goal that each had failed to achieve in intensive conventional therapy. In each case, the (second generation) software could be produced in a clinically appropriate timeframe. Given each patient's previous failure in the goal, a one-two month period to train therapists, treat the patients, and achieve each patient's goal was considered optimistic. The speed with which patients achieved their goal made it possible to expand the study to address additional goals, from the original three to achieving eight. The three case studies make a strong case for the premise that the user interface is the primary design area for cognitive prosthetic software.

This small-scale study also showed a diversity of everyday activities and cognitive deficits that can be addressed by the underlying functionality in calendar software. The second generation software needed some additional modules to deliver the total functionality. The fourth generation software suite makes added functionality much easier to deliver to a patient.

4.3   OTHER EXAMPLES OF THE IMPORTANCE OF UI DESIGN

The primacy of the UI for individuals with cognitive disabilities from brain injury is demonstrated in some additional case studies. The primacy of the UI comes from the detailed ways the UI implements application functionality.

4.3.1   JERRY

Jerry was a victim of a mugging that left him with cognitive deficits, wheelchair-bound, and mute. His cognitive deficits included problems of memory, sequencing, and generalization. He was able to make hand and head gestures. Jerry's family life was complicated, and he had very limited financial resources or ongoing healthcare coverage. He lived with his mother and step-father, a minister. His mother would not turn her adult paraplegic mute son out of the house, but she was not an ally; his stepfather was hostile. Jerry was socially adept, well liked, active in the community including the state brain injury association, and highly motivated for independence. He was on a years-long waiting list for accessible public housing. He lived about 800 miles from the Institute's office. He could keyboard with one hand, but there wasn't a computer in his home. A computer was shipped to the local brain injury association, which arranged the installation of the computer and broadband communications.

Because he was mute, communication was an obvious and important therapy goal. Communication between him and his therapist would need to be primarily through keyboarding. The initial intervention in his case involved writing to communicate with his therapist, which would also be a gateway to communicating with others. Jerry would type using word processing software, and the therapist would speak her reply through videoconferencing. However, she anticipated that shortly he would want to write a journal that needed to be private. The computer was provisioned with a Windows log-on password and a word processor with a single document but with Print, Save, and Exit commands on a drop-down File menu. This served the needs of the first therapy session, since his current computing capabilities were unknown.

In getting to know Jerry, socializing and communicating seemed to be priorities he could control; getting into public accessible housing was highest on his list, but that wasn't something that therapy could affect. Communication was. He had socialized with others playing online chess. It was also hoped that Jerry would now be able to communicate more with his mother.

Within days, it became apparent that Jerry wasn't able to communicate with his mother as she was illiterate, and unable to read his typing. Text to speech was quickly activated and personalized so that they could communicate. The command he wanted was Read, meaning read aloud. A secondary consequence of the text to speech capability was that Jerry became independent in ordering public transportation services for people with disabilities, and also ordering his own durable medical supplies. Before that his mother had ordered both transportation and supplies.

His therapists didn't know how to play chess and Jerry suggested that he would teach a willing therapist. The online chess site he used allowed IM-style communication. This was ideal for him, as it put him on an equal footing with others. He had made some friends in one of the "rooms" on the chess site, and would see them there.

Up to this point, Jerry was using one word processing document, but he came to see the advantage of having multiple documents. So Blank Page was added to the File menu, to indicate a new document in a way that matched Jerry's mental model of the operation.

He also realized that the page of text on the screen was difficult for him to follow when he was using it to converse aloud (with the computer's help), so he asked us to set up one format for conversations, with a hanging indent, in addition to the conventionally indented format to be used with standard word processing documents.

Eventually the drop-down File menu reflected several significant UI changes and functional changes. The Blank Page label was replaced by New, making it correspond to standard UI convention, along with Open, Close, Save, Print, and Exit. On the other hand, the added functionality normally labeled Save As instead received the label Change Name here. Also added was Fax functionality, which is another method for ordering medical supplies to be delivered. Using Fax involved a more complicated procedure and left a paper trail of the order. The text-to-speech commands remained on the menu bar. Each function could be launched with a single mouse click.

He had been a letter writer prior to his injury, and found it useful to type letters to friends rather than handwrite them. He liked the idea of having a contact list of people and organizations. The contact list had a feature that would print an envelope. Next in communication was an email application. This also helped him keep in contact with acquaintances who would become friends.

At this point, Jerry was communicating more and better with friends and acquaintances. He was also able to communicate with his mother. And he and his therapists were communicating by email between sessions. It was decided that therapy sessions were no longer necessary. However, he and his therapists kept in touch via email.

UI design was able to deliver application functionality for Jerry, which in turn gave the therapist the tools she could use to make cognitive rehabilitation successful. His case study highlights a number of points. First, it shows that alternative labels suggested by the software user with deficits can be quite effective. On its face, Change Name is semantically close to Save As, and likewise Blank Page is semantically close to New. But Change Name and Blank Page are both terms that are more concrete and more operational. Because these labels represent the user's own words, they were intuitive to him. These user-customized names reduced the time he needed to learn how to use the commands.

Second, Jerry saw the need to distinguish between keyboarding for conversing with others and creating a document. His suggestion was to mark his conversing with a hanging indentation. This should be considered a contribution to UI design, as it alters how output is displayed to the screen. (The fact that word processors allow the unimpaired user to change the presentation of output is not important in the context of cognitive disabilities, which generally prevent a user from operating at the level of sophistication required to change formats on an as needed basis.)

The ability of the individual to migrate from special command labels for a function to conventional labels is also demonstrated in this example. The patient's own phrasing helps the therapy process and should be viewed as "training wheels" (Carroll and Carrithers, 1984). Therapy is supposed to raise the individual's level of functioning. It is desirable at some point to migrate to conventional interfaces if at all possible, and it is noteworthy that this patient was able to do this.

4.3.2   EILEEN

Eileen was in her 40s and was married with two school-age children. A home contractor's error exposed her to toxic organic solvents. Before the accident, she was active in a family business. She found that she was forgetting tasks and errands, and had headaches as well. Her physician diagnosed her condition as solvent encephalopathy, from the organic solvents. On testing, the problem seemed to be executive functioning and not memory. She had difficulty monitoring her activities, and carrying them out to completion. She also had difficulty in understanding the consequences of her actions and inactions. She was referred by her physician, who had seen the results of 2 other solvent encephalopathy cases she had referred. At this point, Eileen was two years post injury.

During the initial interview, Eileen said that her son's eighth birthday was in three days. On follow up, it seemed that she had invited a dozen extended family members to a birthday party for her young son, who didn't really have a party the previous year. She said that she hadn't shopped for the birthday dinner, or a birthday cake. She hadn't decided on a birthday present or a card. She cared about her son, but was unable to organize herself. This was an example of executive dysfunction in that she was unable to mobilize herself, and felt no pressure to begin working on it. This was also an example of what her life had been like for the previous two years.

It seemed that the Institute's modality of treatment was suited to deal with this problem. A quick survey of the home showed that there was adequate broadband, and a table that could be turned into a desk. It was decided to begin therapy the next day with the delivery and installation of the equipment followed by a therapy session. The Initial Intervention would be to prepare for the birthday present and party. A basic word processor was all that would be needed, one that had a single document plus **Print**. A computer was prepared along with the needed peripherals, and was delivered and installed the next morning. The therapy session used the word processor to plan the party and the present. The therapist helped Eileen identify what needed to be done to organize the party and develop the shopping list. The list was quite specific and identified which store would have what items. They arranged to keep in touch through Eileen's cell phone. The list was sufficiently specific that Eileen had no difficulty with it, and bought what was needed and ordered the birthday cake. The following day was the birthday party, and the minimal food preparation. Again, the therapist and Eileen kept in touch until all was complete. The birthday party was a success in part because a therapist would give her reminders and check in with her several times during the day, when many things needed to be done in a short period of time.

The impact of that Initial Intervention on Eileen and the family cannot be overstated. The birthday party had been heading for another disaster. Then Eileen's new therapy began, with the presence of the computer equipment for all to see. She had moved from having invited people to being prepared for the event; previous events had occurred where there was no preparation. The birthday boy was delighted, as was the family, as was Eileen. The Institute had mobilized to assist Eileen in a way that was different from previous therapy, and Eileen had responded. This therapeutic relationship was beginning on a high note, and we had learned from each other. It was clear that she was able to respond to structure, which would be provided by the therapist on the one hand, and scheduling software on the other.

It was August, and school would begin very shortly. Eileen agreed that the next priority was to get her children ready for school. That would be the focus of the next few weeks, and would involve shopping for clothes, school supplies, pediatrician appointments, filling out forms, arranging for car pools, etc. This involved executive functioning skills that she lacked. It was time to introduce scheduling software, which would be used with daily therapy sessions.

Her therapist introduced the daily scheduling application, to provide structure and daily guidance. Telephone reminders seemed to work for her. The Institute's CellMinder application allows a person to give herself reminders in her own voice—keyed to the schedule—during the day. The combination of the printed schedule coupled with CellMinder became a powerful set of tools for staying on track. Therapy involved asking Eileen questions to help her think through what needed to be done, in varying levels of detail. The word processer would be used for listing what needed to be done, and identifying tasks. From there, tasks would be placed in a daily schedule, which she would print out. Then she realized that posting the daily schedule by the dining area would let others see what they needed to do, and when they were needed. She then realized that she would find it easier to associate a color with each person. She asked that schedule items have a field for color boxes indicating the family member(s) involved in each schedule item (Figure 4.10). She felt that color helped her better process who in the family would be involved in whatever needed to be done, especially when two or more people were involved.

At first she wanted to keep all of her To Do and schedule items. After a couple of months, she mentioned that for the first time, she had become comfortable deleting completed and old items. She was accomplishing things. She also noted that she felt more productive having therapy at home, that when she went to the clinic, she was always rushing to unpack and take notes, and repack, and that by the time she got home, she felt like she was a different person and in a different place. This therapy allowed her to have a normal conversation with the therapist, and also made it easy to work after the therapy session had ended. She also felt that she was able to take more control of things in her life this way. She was also creating little "systems" that involved more regularly scheduled items which related to her own schedule.

The computer software was enabling her to continue to take greater control of her life, and it seemed that her cognitive abilities in executive functioning were increasing. The therapy and the technology were also providing her with support. And she had become very active and very engaged.

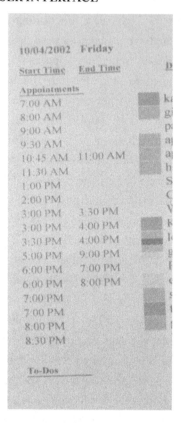

Figure 4.10: Colors assigned to Eileen's family members as displayed in one of her daily schedules. The color coding allowed her to quickly identify the family members scheduled to participate in each activity during the day.

4.4   THE CONTRASTING CASE OF READING DISABILITY

In this case study, customization of the user interface is unnecessary for using commands. This case study represents a group of users who have excellent computing skills in personalizing system resources and computing applications. This group's dependence on the user interface is for reading schoolwork. It is an anomaly that these individuals have difficulty reading, but somehow are able to read documentation on how to personalize their computer. Individuals with learning disabilities were seen as having deficits that computation could bridge, thus enabling the individuals to use their cognitive abilities on subsequent subtasks. A reading disability creates a barrier, impairing progress in learning, making the degree of intellectual disability seem greater than it actually is because the individual is unable to demonstrate abilities on downstream tasks. Computation could provide both functional support and UI support, and there is a sizable commercial market for reading disabilities software.

4.4.1   ZACK

Zack was a high school sophomore with excellent computer-use skills and in special education since first grade. His deficits included visual processing problems, aural processing difficulty, language impairment, and slowed information processing. His IQ had tested in the 65–75 range. He had a fifth Grade reading level, and he was assigned to the Resource Room, the lowest track for classroom teaching in his high school, for all academic subjects. However, several evaluations noted that his abilities seemed to be greater than the quantitative measures indicated. He passed his driver's license test on the second try. He was progressing well in Tae Kwan Do. And he had a power user's knowledge of Windows XP, which he demonstrated during his intake evaluation.

This pattern of demonstrated weaknesses coupled with demonstrated strengths is one that we have always found interesting and promising. How could Zack be able to read at only a fifth grade level but able to read, integrate, and apply information on the functioning of his computer? A speech therapist with a specialty in learning disabilities was enlisted to be his therapist for this project. Because of Zack's computer competencies, he was able to perform user support on issues that arose on his computer and indeed was able to provide user support to the therapist.

Dyslexia—reading disability—is very well studied. In dyslexia, decoding is often a problem, which means that the brain isn't able to quickly process the eye's signals of text. In Zack's case, the effort expended in trying to decode words in a sentence detracted from being able to follow and understand the sentence. Our theory was, though, that perhaps if the text were formatted differently, the brain's difficulty in recognizing letters could be reduced; this approach had been applied successfully in an individual with aphasia (see Jane's case study in Chapter 8). Variables selected were font family, font size, serifs or their lack, spacing between letters, spacing between words, spacing between sentences, and spacing between lines. Newell and colleagues (Newell, 2011) had found that manipulating foreground and background colors improved reading by dyslexics. It is also known that multimodal presentation—print and speech—can aid learning. We decided to use text-to-speech, selecting units from words or sentences or perhaps paragraphs and highlighting each unit as it was read aloud. A reformatting application was built around these variables and goals.

Zack's proficiency in computer use made him very different from TBI patients. Because of his proficiency, he was shown how to use the Institute's personalization tool so that he could customize his own UI. He quickly learned how to use the tool to develop the characteristics that made it easier for him to read a phrase. There was then a substantial increase in his reading speed.

The speech therapist, with access to software developers, wanted to have a word processor with two windows, one for the target text and one for questions and answers about the text or the individual's notes on the target text. The student's school text would need to be scanned in and then sent through the reformatting application to conform to the characteristics selected by the student. Other text, such as the one in Figure 4.11, could be downloaded from the web and then sent through the reformatting application as well. Some special features needed to be added to functionality. It was desirable to save the pair of files as a single unit. This would allow several assignments to use the same source text. It would also be necessary to open the pair of files as a unit.

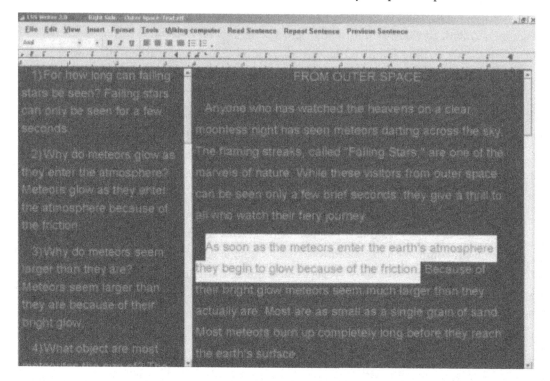

Figure 4.11: Cognitive interventions for reading, embedded in computer software.

The application substantially increased Zack's decoding ability.

Then the speech therapist started teaching the student structural concepts of reading taught early in elementary school: the title or headline of a story gives an overview of the story; the first sentence of a paragraph provides an overview of the paragraph. Zack had a lot to absorb.

The student's positive response to the technology was remarkable. He had been answering about half of the questions in handouts, and then only in short phrases or single words. After two months with this technology, he was answering all of the questions, and in sentences.

His reading developed dramatically over a period of two-and-a-half months. See Figure 4.12 for his baseline performance. Figure 4.13 shows his performance at two months on the same material. And Figure 4.11 shows the 2-window word processor with a science article, and his answers to it; notice the sentences and answers to these questions, again compared to the baseline of Figure 4.12.

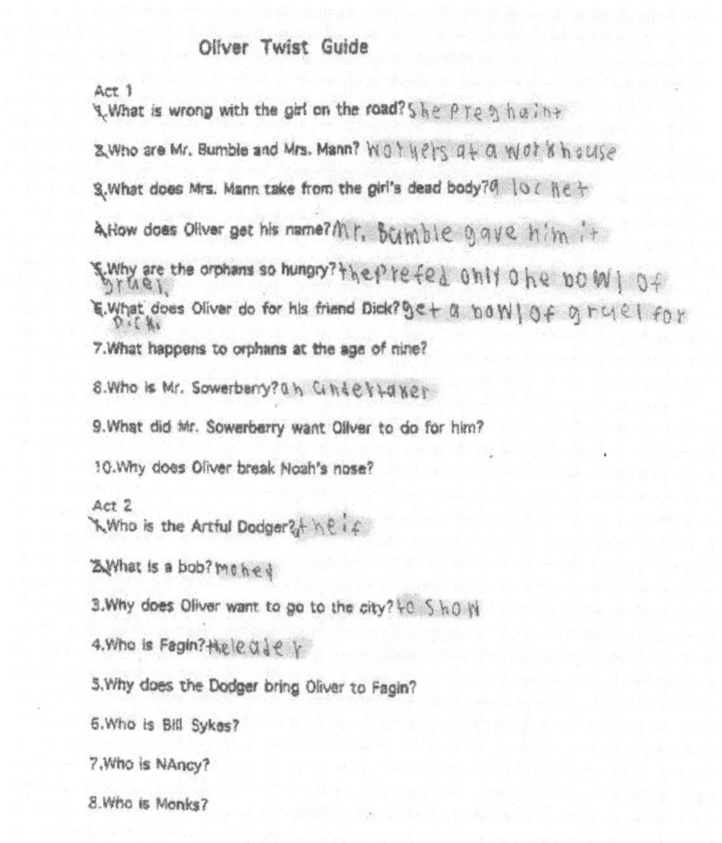

Oliver Twist Guide

Act 1

1.What is wrong with the girl on the road? She Preghaint

2.Who are Mr. Bumble and Mrs. Mann? Wothers at a workhouse

3.What does Mrs. Mann take from the girl's dead body? a locket

4.How does Oliver get his name? Mr. Bumble gave him it

5.Why are the orphans so hungry? the prefed only ohe bowl of gruel.

6.What does Oliver do for his friend Dick? get a bowl of gruel for Dick.

7.What happens to orphans at the age of nine?

8.Who is Mr. Sowerberry? ah undertaker

9.What did Mr. Sowerberry want Oliver to do for him?

10.Why does Oliver break Noah's nose?

Act 2

1.Who is the Artful Dodger? theif

2.What is a bob? mohey

3.Why does Oliver want to go to the city? to show

4.Who is Fagin? the leader

5.Why does the Dodger bring Oliver to Fagin?

6.Who is Bill Sykes?

7.Who is NAncy?

8.Who is Monks?

Figure 4.12: Baseline reading comprehension of an individual with reading disabilities.

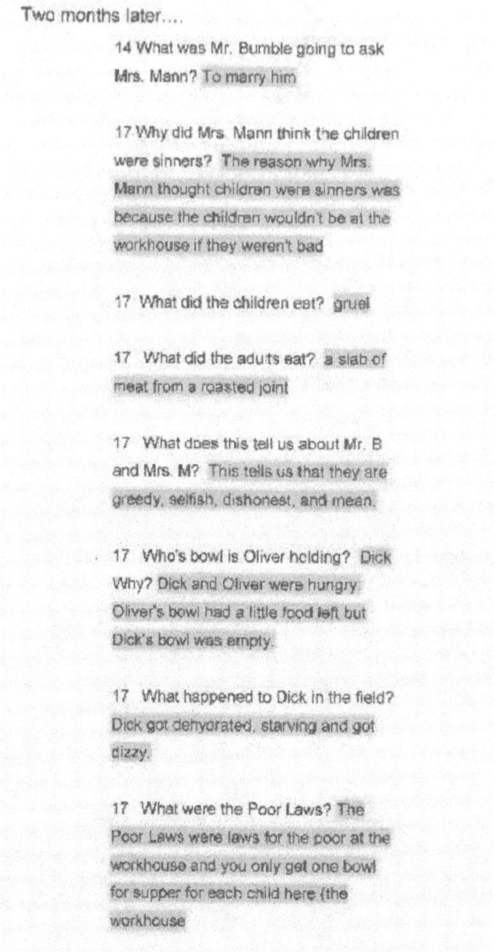

Two months later....

14 What was Mr. Bumble going to ask Mrs. Mann? To marry him

17 Why did Mrs. Mann think the children were sinners? The reason why Mrs. Mann thought children were sinners was because the children wouldn't be at the workhouse if they weren't bad

17 What did the children eat? gruel

17 What did the adults eat? a slab of meat from a roasted joint

17 What does this tell us about Mr. B and Mrs. M? This tells us that they are greedy, selfish, dishonest, and mean.

17 Who's bowl is Oliver holding? Dick Why? Dick and Oliver were hungry. Oliver's bowl had a little food left but Dick's bowl was empty.

17 What happened to Dick in the field? Dick got dehydrated, starving and got dizzy.

17 What were the Poor Laws? The Poor Laws were laws for the poor at the workhouse and you only get one bowl for supper for each child here (the workhouse

Figure 4.13: reading comprehension on the same book after two months of therapy.

4.5 DISPROPORTIONATE RESPONSES TO SLIGHT CHANGES IN UIS FOR BRAIN INJURY USERS

Users with cognitive disabilities require a much better fit for their UIs than do members of the general population, but some seemingly major changes in the UI may show no significant impact on the cognitively disabled user's UI performance. This is because, as large as a change is, it still might not lie in the acceptable zone for the individual. Conversely, a slight change in the UI—one that would be insignificant in designing for the general population—can make the difference between a UI that is usable and one that is useless to a cognitively disabled user.

Standard UI design guidelines offer little support for individuals with substantial cognitive disabilities. With Universe-of-One populations like the TBI patient population, it is likely that substantial UI personalization is needed to make the UI appropriate for the specific individual.

4.6   SUMMARY

This chapter has shown that the user interface is the principal design area for individuals with cognitive disabilities from TBI. HCI has been concerned about designing access to application functionality, and the UI provides access to that functionality. A model of UI performance has shown why individuals with cognitive disabilities need a better fit with the UI in order to successfully use software. Empirical results come from examples of UIs redesigned by patients for their cognitive prosthetic software. These examples showed how small changes in the UI helped make the UI intuitive to the patients, so that training time collapsed to minutes. Patient UI design involved working at a detailed level within each patient's own system. It was often necessary to rephrase commands and instructions to produce a better UI fit. Other techniques involved the use of color, sometimes to soothe and sometimes to provide cognitive arousal. In addition, music was used as a means of engaging the user. Examples were also given of migrating from a personalized UI to industry standard UIs.

CHAPTER 5

Patient-Centered Design

This chapter explores patient-centered design (PCD) as a methodology for personalizing cognitive prosthetic software used as a therapy tool. The history of cognitive assistive technology is that it has been used by patients and caregivers after completion of conventional therapy. That was the original concept underlying the Institute for Cognitive Prosthetics' CAT. But therapists saw the potential of CAT as a therapy tool that could increase rehabilitation gains, i.e., the activities patients can perform. This book documents the realization of some of that potential.

The patient-centered design methodology serves scenarios in which there is active therapy directed at cognitive disabilities. In this approach, clinical priorities are factored into what have only been technical software design decisions. PCD places the user in a clinical context, with its focus on the individual, nuances of the patient's condition, and prospects for treatment and recovery.

5.1   THE PCD MODEL IN THE CONTEXT OF BRAIN INJURY

In patient-centered design, the user is seen in a clinical context. In the case of brain injury rehabilitation, that context is the therapy session. Cognitive assistive technology is used to support a patient-centered activity that will be the content of one or more therapy sessions. Patient-centered design is the methodology that is useful for the design of the cognitive assistive technology that supports the therapy session and the patient's actual activities.

To be useful in rehabilitation settings, cognitive prosthetic software needs to take a small amount of time to prepare, a small amount of time (minutes) for the patient to learn, and must be more effective than current cognitive rehabilitation techniques. The activity for a therapy session is narrow, which means that the software functionality required is very limited, and its UI is small, minimizing the amount of patient learning coupled with UI design needed. This is the basis of PCD serving the needs of the clinical context. The purpose of cognitive prosthetic software is to support the activities of each therapy session, and not to train the individual in the functionality of a device or an application, e.g., word processors or scheduling software or mobile devices.

The therapy session addresses an *instance* of an activity often in the context of an upcoming event. With this contextual information, the therapy task is well defined, making it easier for the patient to process. Patient success with the target activity is entirely focused on the limited and relatively well-defined instance of the activity. The short-term therapy goal is achieved by having the patient be able to perform that instance of an activity.

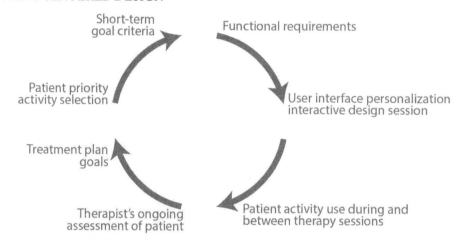

Figure 5.1: Steps in the Patient-Centered Design methodology.

Figure 5.1 outlines the steps in the patient-centered design methodology. The process begins with an evaluation leading to a treatment plan. Clinical evaluations consist of an interview and clinical testing. Additionally, with the Institute's therapy model there is a home visit. The patient-centered design home visit is a guided tour by the patient, with encouragement and interest by the therapist. The therapist gets an opportunity to see the patient at home and observe strengths that may not show in clinical testing. The therapist builds a list of patient priority activities, particularly those linked to upcoming events, and explores the problems the patient has in failing to perform the priority activities with out active caregiver support. This task is made easier by the broad array of artifacts that a visitor sees in a home, and result in questions to the patient, family members, and caregivers. This gives the therapist an opportunity to explore how priority activities are performed. S/he will then evaluate the activities for clinical appropriateness. Then a treatment plan will be written, with long-term clinical goals, and their component short-term goals.

The second step is an analysis of the patient-priority activity with the patient, and determining the Initial Intervention. This will be an instance of the activity, one linked to an upcoming event. This single instance of an activity bounds it and makes it more concrete. An important characteristic of the Initial Intervention is its likelihood of success in a week or two of therapy.

Next is the Short-Term Goal criteria that the patient must achieve for the therapist to declare the goal met. This consists of activity "deliverables," which are tangible work products or observable behaviors. These are artifacts of cognitive functioning.

The Short-Term Goal criteria, coupled with the results of the failure analysis are used in developing functional requirements. Specifying the functional requirements is performed by the therapist. Since the goal is tied to an instance of an activity, functional requirements are simpler to specify, and needn't involve a host of contingencies and options for the generic activity. Additional instances would become one or more Short Term Goals. PCD takes a special approach to subtasks that the individual can perform: the functional design generally excludes tasks and subtasks that the individual can already perform. A disability in the face of an activity means that an individual can't perform the activity without someone's help. In this context, a patient generally experiences a feeling of success in being able to perform subtasks, even if a computer could perform them faster and more accurately.

User interface design is done on the patient's computer, based on the functional requirements. The user interface is the means by which a user is able to gain access to software functionality. With the cognitive prosthetic software suite plus customization module, the UI design can be accomplished in an interactive session with the patient. The patient provides instructions for making the UI intuitive to her/himself, and easy to pass from one step to another. The therapist will likely provide structure for the patient in personalizing the UI. The patient is the best "meter" for his/her UI, and can produce efficient designs. The customization module is also the means of turning on snippets of functionality that will be called by the UI.

After the completion of the UI design, the patient will use the cognitive prosthetic software during and between therapy sessions and therapists will provide substantial guidance. Technology changes process, and the patient will be exploring the new process, and will gain proficiency. During the actual use of the software, the patient may decide that some aspects of the UI design or software functionality needs further personalization. Software designers have long known that some design decisions made during testing have flaws that become apparent when used under actual conditions. Modifications can often be easily done, and the patient is encouraged to suggest changes that are directly tied to a specific subtask. This can lead to additional Short Term Goals. The objective of PCD is not to train the patient to use a particular kind of application such as a word processor. Rather it is to provide the patient with a technological scaffold needed to achieve success in a bounded activity (O'Neill et al., 2010).

The therapist observes the patient working toward achieving the goal. From the Short Term Goal, the therapist can make an objective determination of when the patient has successfully achieved the goal.

PCD is highly iterative. After completing a Short Term Goal, the process begins again. Each time, the therapist talks with the patient to develop a current list of priority activities, selecting one that is clinically appropriate. In PCD, clinical factors drive software functionality.

Clinicians will need training in some facets of PCD. Although they have training in task analysis, the analysis is typically at a coarser granularity than is needed for software user requirements. Clinicians will also need training in the functionality available across the applications in the suite, and in the use of the customization application. Additionally, clinicians will likely need training in the mapping of an activity onto software functionality. Therapists can learn how to conduct therapy sessions and do patient evaluations in two days of training.

5.2 CASE STUDIES

These two case studies demonstrate two very different but effective PCD software interventions and why they are effective

5.2.1 CASE STUDY 1: ESSENCE OF TEXT EDITOR

Bill

The patient was high functioning, competitively employed, and looking for another job. Before his accident, he had been a professional in a government agency, and he knew how to touch-type. He continued to have some memory problems—particularly with his personal and work schedules—as well as some cognitive rigidity, and he displayed reduced frustration tolerance, which could lead to outbursts. He also had dysgraphia—handwriting problems—and pressed down on his pencil hard enough for the words to be visible through several sheets of paper. He was trying to write a short cover letter for his resume to respond to a help wanted advertisement. He would write a draft, want to edit it, and tear the paper as he tried to erase words; an outburst would ensue.

User requirements. The immediate task was to compose, edit, and print a single cover letter. The functionality needed was default font, line spacing, margins, and a Print command. It was necessary for him to be able to insert and delete words, both of which require cursor control. Inserting words could be accomplished by keeping the keyboard in Insert mode. Deleting words could be accomplished by backspacing.

User-interface design. It was also proposed to implement the Print command with a single, dedicated, color-coded function key. Feedback to the user would be in the form of a text box that would appear on the screen. Design of the function-key Print command, and the confirmation text box, would be done by the user in the unstructured part of usability testing. The results of the design session would be added to the application, which would then undergo another cycle of usability testing.

Results. User training took about ten minutes and focused mainly on cursor control for deleting and inserting text.

This intervention was effective. The therapist worked with the patient in creating and revising the letter. The patient was able to make inserts and deletions as he wanted, and he was able to print the product. He kept a hard copy of the letter for his files. He went on to write additional letters, using the previous letter as a template and printing out hard copies. Shortly thereafter, he asked to have a multiple-document text editor, which also required customized save and retrieve features and a customized interface.

This application served clinically important goals and was a major success, even though its functionality was extremely limited. It served the patient's initial and important goal and thus is an excellent example of PCD. Note that the patient solved the problem of how to write additional letters by deleting and inserting. From a software design perspective, the patient's approach would be considered inadequate by most conventional standards, but as a therapy tool for that patient, it was excellent.

This application also made the patient active in the rehabilitation process. It addressed what he saw as a priority activity. He helped design the UI, and he made requests for added functionality. His comments revealed his pride in, and his sense of ownership of, the application and system. That application served as a gateway to other applications that increased the individual's self-sufficiency.

5.2.2 CASE STUDY 2: PATIENT AND THERAPIST CONTRIBUTIONS TO DESIGNING AN ENHANCEMENT TO SCHEDULING SOFTWARE APPLICATION

Beth

This college senior suffered a traumatic brain injury when she was struck by an automobile. Additionally, she had serious medical complications that further reduced her cognitive abilities. She had cognitive rehabilitation while an inpatient. On discharge, it was anticipated that she would need permanent daily caregiver support, and work in a sheltered workshop was anticipated. Consequently, return to college was out of the question. She has a very supportive family. Beth needed complex reminders, and a computer-based calendar combined with a daily printed calendar wasn't sufficient. These were supplemented by the mother, who gave her timely reminders. Well-meaning reminders. But the patient wanted more independence coupled with self-sufficiency.

User requirements 1. The user desired to have greater self-sufficient compliance with tasks that needed to be completed during the day. In particular, she wanted to receive a timely message, based on scheduler items, which would be sent to a mobile device. An alphanumeric pager service was seen as being able to deliver the timely reminder. The message would be an optional addition to a schedule item.

A pager enhancement to the scheduler had been used with other patients in the early days of alphanumeric pagers (see Dr. K.). Because pager companies did not have the ability to store the message and send it at a future time, it had been necessary to build a store and forward message server for delivery of text to the pager service.

User-interface design 1. The user would need to decide if a schedule item needed a reminder, and if so, enter the message. The user would be asked to compose a message for the item, and if none were entered, the item label would be used. The user needed the ability to later add or remove both the option and the content of the message. This functionality had already been designed, and interface customization had already been added to the customization application.

Results 1. The UI was designed and tested, and the solution found unacceptable. A failure analysis revealed two fatal problems. First, the pager's UI was too complicated for the patient, and was not customizable. The then-current generation of pager had multiple modes, and the user was unable to learn how to use any of several devices that were tested. Second, the pager service was unreliable in message delivery, and some messages had arrived a half-hour or more late.

User Requirements 2. The concept of having the patient send herself a reminder remained attractive, and the idea of sending a message to a cell phone was raised. Cell phone calls go through immediately, and delays are in seconds. Delivery failures were signaled immediately. Also, cell phones made it possible for the reminder to be *in her own voice*, so she would be reminding *herself*, a distinct advantage. Enhanced application functionality would include recording start, recording end, playback, and message acceptance. Both Beth and the therapist were enthusiastic about the concept, and the original goal could be kept, with a change in the technology used to achieve it. There was only one problem: There were no companies that had audio message store-and-forward service for telephones. The pager-message store and forward server had some of the functionality required, and would be enhanced to interact with the more complex cell phone devices.

A second set of requirements involved the cell-phone. The phone-server connection is under a protocol and sends messages to each device about the other's status, and this provided for interactive dialogues, e.g., cell-phone answers. Beth wanted to be able to press a button to begin playing the message, and to also be able to repeat playing the message.

User-interface Design 2. There would be two sets of interfaces, one for the scheduler application, and one for the cell phone. The scheduler module already had been enhanced to accommodate a pager reminder message. The patient creates the message in the form (Figure 5.2), which would then be sent to the message-sending application. Beth, like others patients, took some time creating and editing the message, so that it would contain the needed details. Patients seemed to be able to anticipate what details she would need at the time the message was sent. If she did not provide a message, data could be copied from the appointment fields and a sound file created by text-to-speech functionality. The form would also need enhancement and it was anticipated that the form layout would need to be customizable, along with the labels for the commands.

A second set of interface issues involved the cell phone. Beth selected a number that she would press to have the server send her audio reminder; pressing that number would resend the reminder.

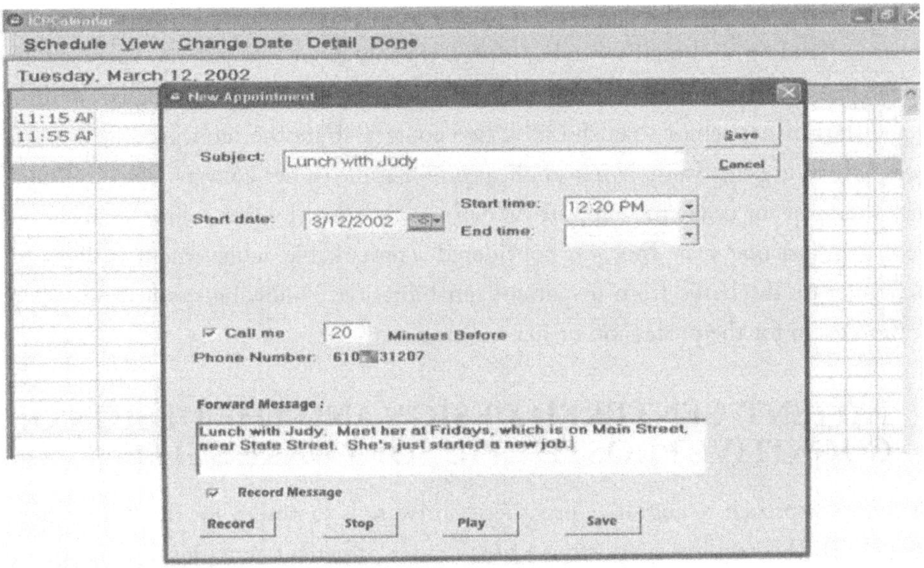

Figure 5.2: CellMinder message form integrated into appointment application.

Results 2. Testing the enhancement involved the two sets of interfaces: The advantages of a scripted message, both in terms of details, and in terms of message fluency in the UI testing. During that session, Beth and her therapist customized the UI for labels and instructions, which were rather conventional. However, when it came to making the recording, some issues emerged from listening to them. There was an uncomfortable pause between pressing the Record button and beginning to speak, and another between stopping talking and ending the message. It was decided to treat the pauses as a user interface issue, because it was burdensome for Beth to begin reading when she clicked the record button, and press stop at the end. Rather, the leading and trailing pauses were automatically removed.

Two issues arose concerning the cell-phone. During testing/design, it was decided to assign a ringtone to messages from the server, so that she will know that the message is a reminder. The second issue involved a delay between the time Beth pressed the button to play the message, and the time the message started to play; she needed a few seconds to move the phone to her ear.

5.2.3   EPILOGUE

CellMinder was a pivotal enhancement. Coupled with therapy, it led to greatly increasing Beth's ability to perform tasks on time. She took great pleasure in being reminded in her own voice. Over time, her therapist saw increases in Beth's cognitive abilities. Her improvement was substantial enough that her therapist felt that a return to college should be explored.

Beth's college responded that she would need to take five competency exams in order to be readmitted. Beth and her therapist designed an application that would be used for relearning the required content concepts. The application "Study Guide" incorporated three modalities of learning—hearing, reading, and tactile (keyboarding). It took about two years to successfully complete the exams. She was readmitted to her senior year of college to begin in August. With a semester before the start of her senior year, she took two courses at another college, and lived off campus. Beth was able to use the Study Notes application for some of her courses. She continued working with her therapist for academic support. When she returned for her senior year, she was able to complete it in just one year. This was considered a remarkable achievement in light of her level of functioning on discharge from in-patient rehabilitation. Following graduation, she passed the certification exam for the profession of her training.

5.3 PATIENT-CENTERED DESIGN AND THE DESIGN OF A COGNITIVE PROSTHETIC SOFTWARE SUITE

The Institute's approach to cognitive prosthetic software is to design for the diffuse cognitive disabilities of TBI. This requires supporting a broad range of patient activities. But another important criterion is that the suite needs to be manageable by therapists, who need to learn the functionality available. Software bloat—having a very broad functionality—makes an application more difficult to learn and remember.

There are two strategies for the therapist to address this problem. First, the therapist solicits a set of activities which are priorities to the patient. Activities that require complicated functionality can be deferred for a current short-term goal. This is the therapist's safety valve.

The second strategy is to look at how available functionality can be adapted to address a patient priority activity. In our experience, a broad range of activities can be supported by a narrower range of applications. For example, Sarah's need for prioritizing lists could be addressed with the time functionality of the scheduling software.

Our application development has been bottom-up, based on the concrete needs of patients and therapists. The cognitive disabilities domain is one that is not well understood, and builds up functionality based on user experience and actual user needs. Encouraging therapists to discuss functional enhancements provides an opportunity to show how existing functionality can be used, when appropriate. It also provides an opportunity to design the enhancement with the therapist; and recent tools make development reasonably easy to do, as in the case of Beth.

5.4 QUALITY IMPROVEMENT STUDIES VERSUS MEDICAL RESEARCH

The Quality Improvement (QI) model (Langley et al., 2009; Ogrinc et al., 2008) can be used with PCD, and provides a means of intervention improvement over the typically months-long treatment for each patient. This has substantial advantages over the Medical Research model, which freezes the intervention for the duration of the study. QI provides an opportunity to learn "what works" for a large number of Universe of One users. The resulting cutting edge improvements can then be examined for patterns, resulting in a faster pace of empirical advancement.

Quality improvement studies have a different set of assumptions and standards. It is directed toward improvement in diagnosis and/or treatment of *direct benefit* to the patient, and it presents no more than minimal risk to the patient. The focus is improving clinical processes in an ongoing process of plan, do, study, and act. The patient treatment generally follows a standard treatment but with streamlined processes of care. In QI studies, there is a focus on benefit to the specific patient, maintaining the ethical commitment to serve the patient's best interests and treatment preferences. Results of the Quality Improvement projects can be published in journals and related websites and presented as a Case Report. Patient confidentiality is protected as it is in clinical studies.

In medical research studies, in contrast, it is assumed that a research subject will not necessarily receive any direct benefit from the proposed intervention because of randomization and placebo controls. Furthermore, research subjects may well be exposed to greater risk and substantial dangers in the proposed research. Cancer research provides an excellent example. An experimental chemotherapy drug might be undergoing trials for its ability to increase the survival of patients with a particular kind of cancer. The research is likely to be invasive, as when a drug is administered to the patient. An administered drug is likely to have some harmful side effects, which may or may not be greater than the side effects of the alternatives. Research subjects—patients who have been accepted into the study—are randomly assigned to the group receiving the experimental drug or to the control group. In the classic double-blind study, clinical staff members treating study subjects are unaware of who is receiving what treatment, so no one can project expectations. At the end of the study, the researchers continue to collect and analyze data. Even if a subject-patient has responded very well to an experimental drug, s/he may be denied post-study access to the drug. Clinical trials are needed to evaluate an experimental drug for both its safety and its effectiveness. The informed consent process, Institutional Review Boards, and Safety Monitoring Boards are required to protect research subjects.

The PCD model is strongly patient centered and noninvasive and may be viewed as quality improvement rather than conventional medical research. Therapy incorporates the patient's priority activities, particularly those tied to upcoming events. Therapy interventions rely initially on conventional techniques but then can be enhanced with technology to enable the patient to have a greater recovery of function in areas that are important to the patient. Technological scaffolds are highly customized to the needs of the individual patient. The patient receives therapy at home or his/her other settings. This eliminates the cognitive load and disadvantages of receiving therapy in a clinic. A cognitive prosthetic software suite is used in CPT therapy, and new features are added to the suite in response to patient needs.

In the QI model, therapists and IT staff are free to make changes—and should be encouraged to make changes—that can improve the progress and outcome of the patient. With today's technology, changes can be made in a time frame appropriate to the case. Enhancements are designed for each specific patient. However, it is anticipated that enhancements stimulated by one patient are likely to be adaptable to be useful to other patients as well.

QI activities may need to be supervised by the institutional review boards if there is a significant modification of the standard treatment process. However, QI activities are likely to be eligible for expedited review.

5.5   SUMMARY

Patient-centered design is a software design methodology that is useful for the design of cognitive assistive technology when it is used as a therapy tool. Here the clinical priorities of a therapy sessions are factored into the design of a patient's CAT, which is used as cognitive prosthetic software. PCD is an iterative design methodology that provides the cognitive rehabilitation therapy patient with the functionality s/he needs for a therapy session goal, and the customization of the user interface for that functionality. As the patient achieves a short-term therapy goal, another is selected in a process that involves a patient priority activity that is clinically appropriate at the current point in therapy.

Two case studies are used to illustrate PCD. In the first case, an important short-term goal was achieved using software that truly had minimal functionality. The patient achieved the therapy goal in minutes. The second case study described a technologically complex enhancement to the patient's software. That enhancement was pivotal in enabling the patient to achieve life-changing goals and demonstrate an increase in cognitive abilities. That enhancement became part of the cognitive prosthetic software suite.

This chapter also described Quality Improvement research. QI is particularly well suited for the refinement of cognitive assistive technology to the particular needs of the individual patient. This enables therapists and designers to push the CAT and therapy envelope to build technology tools and delivery system components that produce a better outcome for the patient. It can also help develop higher performance and lower cost systems. QI is appropriate when the research effort is intended to benefit the patient participating in the research activity. This chapter adds to the book's premise that cognitive assistive technology can and should be an important therapy tool, by increasing the patient's cognitive functioning and in some cases suggesting an increase in the patient's underlying cognitive abilities. CAT adds tools for therapy to the tools for living and tools for learning developed by Carmien and Fischer (2005).

CHAPTER 6

Cognitive Prosthetics Telerehabilitation

This chapter discusses Cognitive Prosthetics Telerehabilitation (CPT), a modality of therapy for cognitive disabilities. It basically combines cognitive prosthetic software with distance delivery of services into the patient's setting. CPT is a generic term for technology-based cognitive rehabilitation therapy, not just the specific implementation developed by the Institute for Cognitive Prosthetics. CPT can be implemented in a number of different ways to fit the national and organizational environments in which it is used. Indeed, CPT is an innovation that will be modified by complex social environments (Rogers, 2003, Chapter 10). CPT has the prospect of being applied to other diseases and conditions producing cognitive disabilities, not just TBI. CPT is an extension and updating of computer-based cognitive prosthetics in our earlier publications.

This chapter begins with a presentation of the elements of a basic model for the CPT modality. Next is a discussion of the cognitive load of in-clinic cognitive rehabilitation services, which has become more and more relevant with the reduction of in-patient cognitive rehabilitation. Following that is a discussion of the implications of eliminating the travel factor in the delivery of cognitive rehabilitation. Then there is a discussion of other advantages of CPT that have become clear in our work with TBI and related acquired brain injuries (ABIs). There is a discussion of the Institute's implementation of CPT and issues unique to the distance delivery of cognitive rehabilitation services (i.e., cognitive telerehabilitation). This chapter ends with a discussion of how clinicians, patients, and organizations can add to, expand upon, and modify the CPT model as they find additional uses for the technology.

6.1   THE COGNITIVE PROSTHETICS TELEREHABILITATION (CPT) MODEL OF COGNITIVE ASSISTIVE TECHNOLOGY

CPT is a form of cognitive assistive technology. Presented here is a generic version of the approach developed and used by the Institute for Cognitive Prosthetics. This model is an outgrowth of our previous models, resulting from advances in technology as well as contributions from therapists and patients. In other words, it has been tested in the field under actual conditions of use in patients' various settings.

- CPT is designed specifically for rehabilitation purposes.
- It is intended initially as a platform for cognitive rehabilitation therapy with potential gains in cognitive functioning; following the completion of cognitive rehabilitation, it can be used as a personal productivity tool.
- The patient receives therapy at home and perhaps in other settings; the therapist delivers therapy from a distant location.
    - The patient's computer has cognitive prosthetic software installed on it.
    - There is an electronic workspace shared by the therapist and patient.
    - The patient and the therapist talk to each other primarily via videoconferencing.
- Therapy addresses an activity that the patient considers high priority.
- The therapist applies technological scaffolds in addition to conventional scaffolds to support patient progress.
    - The technological scaffold is highly personalized to the patient's needs.
    - Software functionality presented to the patient supports problematic subtasks.
    - User interface UI design and redesign are primarily patient tasks .
- A cognitive prosthetic software suite provides functionality, invisible to the patient in the beginning, that can be easily invoked to support a broad range of the patient's potential activities. The suite makes the software appropriate for cognitive rehabilitation services.
- The technology collects clinically valuable usage data.
- CPT takes advantage of the elements of patient-centered design. Software is customized to the needs of the patient.

Figure 6.1: Elements of the cognitive prosthetics telerehabilitation model.

The CPT model outlined above contains two important additions to the computer-based cognitive prosthetics model presented in Cole, (1999). The first is the explicit recognition that CPT is to be used as a platform for cognitive rehabilitation therapy. Obviously, CAT that is capable of producing increases in cognitive functioning ought to be used as a therapy tool. However, only a relatively small number of therapists have recognized computer software as an important tool in cognitive rehabilitation (cf. Green[6], 2011; McCall et al., 2009; and Linebarger, 2007).

The second is the recognition that some CAT can promote increases in cognitive abilities in some patients, producing a partial cure, where brain plasticity is a possible mechanism (Theodoros, 2011). (It is likely that some CATs are capable of producing changes in cognitive abilities while others are not.)

The model also combines cognitive prosthetic software and delivery of services in a manner that does not require travel on the part of either the patient or the therapist. It will be shown later in this chapter that this has important implications for the delivery of rehabilitation services.

---

6 Joan Green, MA. CCC-SLP is a long-time advocate of the use of computer software in cognitive rehabilitation. Her company is Innovative Speech Therapy (www.innovativespeech.com).

A patient's settings are places where the patient performs at least some everyday activities. In functional cognitive rehabilitation, these places are important to treatment because their characteristics affect the way patients perform their activities.

The CPT model calls for therapy to be delivered in a patient's setting (usually the home, but the workplace, school, and places in the community are other possibilities). Most patient activities need organization and planning and are conducive to being integrated into cognitive rehabilitation and dealt with in the home. An advantage of CPT is that the therapist can present aspects of planning and organization during a therapy session, leaving the patient to move forward with planning and organization between therapy sessions. Since the patient considers the targeted activities important, s/he will typically want to work on them between sessions. This sets up a pattern of behavior that is a goal of cognitive rehabilitation: actions based on the patient's ability to move activities forward and successfully perform them. This goal can best be accomplished with CPT elements.

It is anticipated that other organizations applying the CPT model can obtain results consistent with those reported in this book.

6.2   THE COGNITIVE LOAD AND OVERHEAD ASSOCIATED WITH TREATING A PATIENT IN THE CLINIC

There is a cognitive load associated with traveling to and from a clinic, and this cognitive load reduces the patient's capacity to receive the full benefit of the therapy services provided (see Theodoros, 2011). Going to and from therapy involves stresses and stressors—and little opportunity to rest.

The patient must get organized for the trip, which includes getting dressed and eating as well as finding and packing materials that will be used in therapy. Often there is time pressure. Other stressors might be the time spent in the vehicle, the jostling of the vehicle, traffic, and weather issues such as rain, sun, heat, and cold. Arriving at the facility, the patient will need to find the way to the clinic. At the clinic, the patient will need to remove outdoor clothing and wait for a therapy session. When called in to therapy, the patient may need to unpack and organize papers that have been brought from home and lay them out in the space provided by the therapist. At the start of therapy, the patient may well already experience some amount of fatigue and not be well rested. Brain injury patients typically have reduced amounts of physical or mental stamina. The impact of being tired produces a much steeper decline in this population's cognitive functioning than experienced by non-injured individuals who happen to be tired.

Cognitive rehabilitation sessions are typically scheduled for one hour. The patient may be able to tolerate the session for that length of time, but the patient's ability to benefit may erode during the session because of the cognitive load before the session on top of the cognitive load associated with the session itself. If the patient is in an intensive program with therapy sessions scheduled back to back, the patient doesn't have the opportunity to lie down and rest between sessions. At the end of a day's therapy, the patient needs to reverse the travel procedure. The patient arrives home and will almost certainly be cognitively (and probably physically) tired.

The cognitive load of in-clinic therapy is rarely considered, let alone quantified. The patient's setting should be considered the more desirable alternative unless the environment is so hostile and disruptive that the patient will do better away.

6.3   THE ADVANTAGES OF CPT

CPT has broad advantages for stakeholders in brain injury rehabilitation. These specialized services become broadly available rather than available in larger cities. Therapy can be made more concrete, more appropriate to the patient's ability to understand the points made in the therapy session. There are advantages in user experience that apply both to the patient and the therapist in treating the patient in the natural setting. Therapy session length can be adjusted to end when the patient's ability to absorb begins to drop. And CPT costs are about 30% less for therapy, not counting the increased patient outcomes. And there are gains when there is a continuity of therapists across the continuum of outpatient care.

6.3.1   THE AVAILABILITY OF SPECIALIZED BRAIN INJURY REHAB SERVICES WHEREVER THERE IS BROADBAND INTERNET SERVICE

Brain injury rehabilitation is a specialty service that has limited availability; in fact, it may be correct to say that it is widely unavailable. In many areas, especially rural areas and small to medium-size cities, the specialized services of brain injury rehabilitation are not available at all. In larger metropolitan areas, although there are typically multiple facilities, a large percentage of patients live more than an hour away, especially when considering rush hour traffic. But even for those who have to travel only a short distance, there is a cognitive load associated with traveling to clinics for cognitive rehabilitation services.

Telerehabilitation therapy services use broadband Internet service. Telerehabilitation services are not particularly high-bandwidth activities, and most Internet subscribers are likely to have access to good service. Broadband Internet service is far more prevalent now than in-clinic TBI rehabilitation services. And broadband Internet services are expected to continue to expand geographically.

Videoconferencing has become commonplace, and there are a number of free services that are widely used for videoconferencing. Telerehabilitation and telemedicine in the US requires HIPPA compliant communications, and requires protection such as encryption, and also provides good to excellent lip synchronization, picture size, and smooth images at a modest cost. Brain injury patients benefit from good communications.

6.3.2 THE ADVANTAGES OF TREATING THE PATIENT IN THEIR OWN SETTING WITH CPT

Treating the patient in the natural setting (Figure 6.2) via telerehabilitation has many advantages over treating the patient in the clinic, and none of the disadvantages of the clinic involving cognitive load. And the patient's own setting is often the best place for an individual recovering from TBI to receive cognitive rehabilitation therapy.

Figure 6.2: A patient in her natural settings in session with a therapist via telecommunication. The patient's computer workspace is shared with the therapist during the session.

A number of issues come into play when an individual receives therapy at home, including where the patient works, the content of therapy, motivation to do homework, patient engagement, and patient stamina.

As has been pointed out earlier, working with the patient in his/her setting, the therapist is able to superimpose therapy on top of the patient's actual activities. If the focus is on patient activities, there is value in focusing on activities the patient considers high priority, particularly ones that are clinically appropriate. Addressing *patient* priority activities not only increases the benefit to the patient, it also promotes patient engagement by serving the patient's agenda. It increases the patient's ability to gain from the therapy because it is more concrete and more salient to the patient. Working with the patient in his/her setting makes it easier for the patient to get artifacts relevant to the event or activity that is the subject of the therapy. The patient better understands the relationship between what is happening in therapy and what is happening in real life.

With telerehabilitation therapy at home, the patient needs a place to receive therapy. That involves a place for the technology, as well as a place to keep the materials needed for therapy. In other words, the patient needs a desk or table dedicated to therapy activities. This space will be used not only for therapy but also as the place where the patient can plan and organize many of the activities in his/her life.

Through telerehabilitation conversations, the therapist gains considerable insight into a patient's detailed performance of activities, and is able to gain growing insight into the patient's life. The therapist is able to directly observe much patient behavior related to in-context activity performance. Patient comments provide descriptions of the activity's performance out of therapist view and allow the therapist to inquire conversationally about more details. The therapist also has a chance to see various artifacts surrounding target activities and events and has the opportunity to inquire conversationally about these as well. In this way, the therapist is able to gain information about, and insight into, the patient's performance. Conversation may also reveal upcoming events that the therapist may want to use as vehicles for therapy. The therapist is able to apply all this contextual information to fine-tuning therapy interventions to better fit the patient.

Between therapy sessions, the patient will likely continue to work on the therapy activities, which after all are patient priorities. And the patient will be able to pick up where the therapy session left off, in contrast to the situation when a patient returns from treatment in a clinic. The therapy activities in this model are not just homework but are a desirable part of the patient's life.

Therapy delivered to the patient's setting eliminates the cognitive load of traveling to and from the clinic. There is no traveling, except to the desk. There is no packing and unpacking of materials. There *is* an opportunity to rest before and after each session. The therapy tasks are qualitatively different from in-clinic tasks, because the in-clinic therapist doesn't have the ability to get into the patient's setting in order to design homework that takes advantage of the patient's high-priority activities, and the in-clinic therapist can't observe the patient's work at home and fine-tune the intervention accordingly.

Again, the goal of the therapy session is to help the patient prepare for and plan for the high-priority activity and event. And this can best be done by the patient sitting at a desk, talking to the therapist, thinking, and typing. (Small-format computers, by the way, with their small screens and small keyboards, are poor vehicles for therapy sessions (Tremaine et al., 2005).)

A Patient's User Experience and a Therapist's User Experience

At a conference on telepractice for speech therapists, the Institute's principle speech therapist was invited to give a presentation. Dr. Sonya Wilt was also the chair of the Pennsylvania Speech Language Pathology Board of Examiners, which licenses and regulates speech therapy practice, and the conference was the annual national conference of state boards regulating speech therapy. The presentation included portions of a telerehabilitation session with formal testing, and work on a patient priority activity. The patient was several years post injury and had several other cognitive rehab experiences that were unable to effectively treat her executive functioning and memory problems. In a video for the conference, she told the attendees about the major differences in her experiences. With conventional therapy, she was rushed in the therapist's office, unpacking materials, struggling to take notes, repacking, and then travelling home where she attempted to remember what had happened and to apply the suggestions. When doing therapy via cognitive prosthetics telerehabilitation the patient found that sitting at her desk was an important element. She felt like an independent person working on a problem with the therapist close enough and far away enough to provide assistance. The therapist had the patient do the keyboarding in the application. That provides the continuity between therapy session and working on her activities between sessions. The patient and therapist were 300 miles away from each other but the therapist was much more involved in their patient's life because the patient and the therapist could deal with activities of the patient's life.

For her part, Dr. Wilt was able to provide guidance to her patient by asking questions for her patient to ponder. Their conversations provided Dr. Wilt with information about upcoming events, the patient's experiences during the day, and life in the household. It was this information that enabled her to have a much better understanding of her patient's life, and how she could better treat her. Patient gains, compared to conventional therapy, also gave Dr. Wilt a much greater sense of satisfaction in helping her patient, and in her profession. Her user experience had been enhanced.

6.3.3   THERAPY SESSIONS THAT FIT THE PATIENT'S ABILITIES AT THE MOMENT: FLEXIBLE LENGTH SESSIONS

Technology changes process. There are designed changes, anticipated changes, unanticipated changes, and forgone changes. With telerehabilitation into the home, neither the patient nor the therapist travels, so neither has a "travel investment" in the length of a therapy session, i.e., neither would feel travel time was wasted if the session only lasted ten minutes.

This has important implications for the relative effectiveness of therapy—of the capacity of the patient to gain from the therapist's efforts. Brain injury patients can have compromised cognitive stamina, which varies from day to day and within a given day. Worse yet, as the individual begins to tire, cognitive performance drops sharply, and along with it the ability to gain from therapy. Telerehabilitation provides the opportunity for therapists to break up therapy into units that improve the patient's ability to gain. With this approach, 60 minutes of therapy delivered in segments during the day should be more effective than a single 60-minute segment.

There are four scenarios that can come into play with cognitive therapy sessions. The first is when the patient becomes cognitively too tired to benefit from therapy. Patient cognitive energy can fluctuate widely during the day. At lower levels, the patient's ability to participate and gain may be inadequate. The patient may lose cognitive stamina during a therapy session. At least sometimes, the patient will be at a low level at the beginning of a therapy session. Therapists try to teach patients to pace themselves; when they become too tired, they should take a nap and recover their cognitive strength. When this happens with in-clinic therapy, it is logistically impractical for the session to end early so that the patient can take a nap. But continuing the session sends the wrong signal to the patient—that the patient should continue even when fatigued. However, telerehabilitation sessions are relatively easy to end and reschedule. (The therapist's schedule will be discussed below.)

In the second scenario, the therapist and patient finish the therapy session early. They have worked on a task and completed it; alternatively, only part of the session's plan may have been achieved, but the therapist feels that the patient needs to reinforce what has been accomplished. At this point, the therapist should suggest an assignment for the patient to work on "independently under supervision," and the therapy session should end. With in-clinic therapy, logistics get in the way. With telerehabilitation, this is easy to implement.

The third scenario involves therapy that is split into two or more sessions for the day. There are several variants here, including a session intentionally split in two, as discussed in the second scenario, because the therapist wants the patient to work independently on a structured task. Another variant would be a session split in two and postponed because of patient fatigue, as in the first scenario. In all the variants, the patient has an opportunity to rest between sessions.

The fourth scenario involves a question or roadblock that arises while a patient works independently on tasks for an activity. Progress on the task at hand stops. With telerehabilitation, however, the problem can be addressed by a brief therapy session focused on the question or roadblock. The patient contacts the therapist directly, requesting help on a problem, and the therapist replies at a convenient time. The therapist is usually able to address the problem in minutes, sometimes solving the problem and other times redirecting the patient and addressing the problem in the next therapy session. The major advantage of this approach is that the problem doesn't fester while the patient waits for the next scheduled therapy session.

In all of these scenarios, telerehabilitation allows the *substance* of therapy to have priority over a fixed *schedule*, allowing the patient to gain more from therapy than would be possible in a clinic. Telerehabilitation also allows the therapist to respond to events that happen during the therapy session, including occasions when it is obvious that the patient is too fatigued to continue. Therapists using telerehabilitation will have more flexible time during the day and will have the ability to treat the same caseload—perhaps even a greater one. Patients generally will experience more flexibility in their schedules as well. In the Institute's experience, therapists and patients have rarely had difficulty arranging mutually agreeable times to connect.

Diffusion of innovations in organizations literature—related to the resistance to change—speaks about the way complex innovations such as telerehabilitation are adapted by organizations to fit organizational preferences. Having flexible length therapy sessions is not a basic component of cognitive prosthetics telerehabilitation. Indeed, one can anticipate that organizations will want to have a substantial amount of experience with CPT before formally introducing flexible sessions, and may decide agains it. On the other hand, one could speculate that some therapists would respond to a patient's request for brief help on a problem before a flexible-session policy was in place.

6.3.4   A DIFFERENT COST AND REVENUE STRUCTURE FOR CPT THERAPY

CPT programs have different cost structures from in-clinic cognitive rehabilitation therapy programs. CPT therapy should show substantial decreases in cost. At the same time, telerehabilitation can allow patients to make faster gains and achieve a more complete recovery.

Patients are better able to *receive* therapy services. The patient sits at his desk and works. The patient's cognitive stamina hasn't been reduced by the cognitive load of travel. The therapy session's length can be geared to the patient's cognitive stamina. The context of CPT therapy is the patient's actual activities and the patient's own priorities. CPT is more relevant to the patient than in-clinic therapy, and the level of therapy intensity is higher with activities important to the patient than with generic activities in a clinic. Therapy session gains are hardened by homework, which is tied to the patient's high-priority upcoming activities. Homework sessions magnify the impact of therapy at no additional cost to the payer. Also, the technology is available for the patient to use with their activities seven days a week.

Moreover, the therapist is more effective when cognitive prosthetic software can be used to bridge cognitive deficits because the therapist can deal directly with the patient's actual activities rather than indirectly with generic activities. Addressing actual activities allows more precise therapy. Actual activities involve situational complications that the therapist can address. As therapy progresses, patients are able to spend increasing amounts of time on their activities between sessions, hardening gains. The ratio of non-contact time to contact time with the therapist increases, further increasing therapist productivity while reducing unit cost of patient services.

The Institute's experience suggests that the effects of the use of telerehabilitation on both the patient and the therapist combine to enable many patients to make faster, more complete recovery than in a clinic. This substantially lowered the cost per unit clinical gain.

There are additional costs associated with CPT therapy, however, including the cost of technology for patient and therapist, the cost of user support and training, and the cost of Internet service. Hardware costs are low and falling. Software costs have substantial volume discounts. Professional support staff ratios are low. All these are substantially offset by therapist productivity gains. On the other hand, there are resource savings as well. A waiting area for patients and an area for clinical therapy aren't needed, nor is staffing for them. There is often additional revenue because patients can be treated on days when they otherwise would be unable to travel to the clinic.

6.3.5   CONTINUITY OF THERAPISTS ALONG THE CONTINUUM OF CARE

Cognitive rehabilitation services are structured into a number of stages, from inpatient programs to day-hospital programs to vocational services, fill-in services, and discharge. Typically, therapists are assigned to particular stages, and often therapy organizations do not serve the full continuum of outpatient services. Once patients complete one stage, they are handed over to other therapists working in the next stage.

Each transition involves a learning process for both the therapist and the patient, most likely greater for the patient. There is also an adjustment process for the patient. In some instances, there are administrative conveniences for the provider organization. The adjustment process that the patient needs to go through is inefficient for the patient.

Telerehabilitation makes it feasible for the patient to keep the same therapist along the entire outpatient continuum of care. This has advantages for both the patient and therapist. The patient doesn't need to adjust to a succession of therapists, who are not readily interchangeable, after all. And the therapist gets to see the progress that the patient is able to make, an experience that is bound to be professionally rewarding.

But there is an additional factor that is important. At the end of therapy the patient is discharged because regular services aren't needed. And most likely, after discharge there will be a crisis that will destabilize the patient. Where there has been a long relationship, and a telerehab relationship, it is clinically and technologically easy for the therapist to intervene and help the patient through the event. In some cases, the amount of time needed to resolve the problem is minutes. In other cases, it would be worthwhile to reopen the case. This can save a substantial investment in therapy services, quickly address threats to clinical success, and have a lasting impact on the former patient.

6.3.6 MOBILE DEVICES

Mobile devices have been touted as providing a new technology generation for individuals with cognitive disabilities. Devices studied have largely been personal digital assistants (PDAs) and to a lesser extent, smartphones. Hundreds of thousands of these units have been distributed to veterans wounded in the Iraq and Afghanistan wars. Therapists were neither trained in the use of these devices nor in how veterans suffering from TBI and concussions could use them.

The focus of this book is on CAT, both for therapy sessions and for planning and performing activities. Mobile devices—smartphones—have been used in CAT as mobile adjuncts to computers with sizable screens and keyboards. As primary devices for therapy, though, current mobile devices have major UI deficiencies that are too difficult for TBI patients to overcome. Smartphone screens are too small to display much information, and keyboards, virtual and chiclet, are too small for patients to use easily.

However, the portability and processing power of smart phones make them ideal as a personal network device connecting to the computer used for therapy. They are very well suited for providing context-aware information to support activities. It is likely the UI issues will drive the network role of current and future portable devices. Current mobile devices are of limited usefulness in cognitive rehab, although their portability is a major advantage. However, advances in smartphone accessories and in tablet computers are resulting in larger units. These most likely will be able to be personalized so that they can serve as therapy platforms. But the UI is the primary design area of interest when it comes to individuals with cognitive disabilities, especially brain injury.

6.4   SUMMARY

Cognitive prosthetics telerehabilitation (CPT) is a generic model of brain injury rehabilitation. CPT transforms this specialty service from limited geographic availability to one that is availably anywhere there is the Internet. Also, CPT has been shown to be superior to conventional in-clinic TBI rehab in many significant ways. There is evidence that patients can have a faster and more complete recovery and at a cost that is less than for in-clinic services. Defining characteristics include delivering service to patients in their own home, having their own priority activities be the focus of rehabilitation, and using cognitive prosthetic software coupled with videoconferencing and sharing workspace. During therapy sessions the patient can prepare, plan, and perform activities in the home and prepare and plan for activities that will be performed elsewhere. Mobile devices can be incorporated to support activities outside the home. CPT begins as a therapy platform. At the end of therapy, the technology becomes personal productivity software and is used independently.

CHAPTER 7

The Active User and the Engaged User

Over the past decade, user engagement has become an important goal in HCI (cf. Norman, 2011). During the same time frame, the rehabilitation area of psychology has devoted attention to the importance of patient engagement as a mediator of rehabilitation success (Lequerica and Kortte, 2010; Danzl et al., 2012). This chapter will elaborate on how users become active and engaged through the use of cognitive prosthetic telerehabilitation, and how the active user and the engaged user are important to CPT, CAT design, and successful patient rehabilitation outcomes.

Assistive technology is an area where device abandonment is often a major issue, sometimes exceeding 50% of units delivered. If one were to explore underutilized devices, the resulting percentage would be even greater (Carmien 2010; Dawe 2006; Scherer 2012).

It is helpful to distinguish between an uninvolved user, an active user, and an engaged user. It is worth noting that a sizable percentage of so-called users are likely to have abandoned the use of their CAT devices. An uninvolved user of a CAT device, assuming s/he is still using the device, will exert minimal effort in the process of using the device, and the device will have consequently provided a low level of value to the user. An active user will have become a self-starter in the use of the device. The active user will be involved in the various activities and processes that are part of the usage process. The engaged user of a CAT device will have a substantial emotional connection to the device and may see it as an extension of himself/herself. Astell et al. (2010) also discuss the idea of making the user active.

7.1 PATIENT PRIORITY ACTIVITIES AS THE FOCUS OF THERAPY

The dominant modality of cognitive rehabilitation therapy is functional rehabilitation, which uses activities as vehicles for addressing cognitive disabilities. The functional rehabilitation model has a good fit with apps and personal productivity tools, which support patient activities. Patient priority activities are almost never the focus of conventional cognitive rehabilitation. Rather, generic activities are selected, and they are typically removed from what the patient experiences in everyday life.

A therapist's initial interaction with a patient involves an evaluation, which can be used to elicit the patient's priorities in terms of activities needing support. As previously discussed, however, in the typical clinical setting, they are generic. A patient-centered approach uses patient priority activities as the vehicles of therapy. As the term implies, patient priority activities are activities that the patient wants to perform independently but hasn't been able to perform without caregiver support since the injury. So in the initial evaluation, the therapist will be able to have the patient express at least some priorities that are likely to be good candidates for the current stage of therapy. This list will change during therapy, and it is important for the therapist to keep up with the change.

A patient who has undergone conventional therapy will experience the therapist's invitation to spell out the patient's own priority activities as an unusual occurrence and a signal that this round of therapy will be different.

7.2 EARLY PRIORITY ACTIVITIES THAT ARE LINKED TO UPCOMING EVENTS IN THE PATIENT'S LIFE

Clinic-based therapists have difficulty learning about the activities of patients who come to the clinic. Patients typically have disabilities in memory and organization and are not particularly reliable informants about their future activities.

In contrast, a patient receiving therapy at home (from a clinic-based therapist) has many information sources available to consult, both people and documents. The therapist will be able to make the patient active in obtaining the information needed. This kind of information will constantly be necessary for planning therapy sessions, and the therapist will constantly be asking the patient about upcoming events. Over time, the patient will become capable of maintaining this information and supplying it to the therapist.

The therapist will likely have options in selecting a priority activity tied to an upcoming event. S/he will be able to evaluate the options and select one that seems the most appropriate, given the stage of the patient's recovery. This is important for the TBI patient, who, because of cognitive deficits, can deal with a concrete activity better than with anything generic.

Again, this approach helps make and maintain an active user of cognitive assistive technology because the patient realizes that the focus of therapy is on events of importance in the individual's own life.

7.3 FAST ACHIEVEMENT OF INITIAL INTERVENTION IN FUNCTIONAL REHABILITATION

The initial intervention is designed to reinforce the active user experience and to create the emotional experience leading to engagement. The initial intervention is designed to allow the patient to rapidly experience success with the new and different treatment modality. Typically, the initial intervention will be one that allows the individual to achieve success within a week or two, in a priority activity that s/he hasn't been able to perform since the injury without caregiver involvement. This success provides rapid and concrete evidence of therapy gains, especially compared to conventional TBI therapy.

Technology use requires preparation. Among the stages of the preparation process are (1) the therapist's visit to the patient's home, (2) the design of the user interface with the active participation of the patient, and (3) the beginning of the therapy sessions.

7.4   THE THERAPIST'S VISIT TO THE PATIENT'S HOME

Part of the CPT therapist's evaluation is a tour of the patient's home. The patient is actively involved in the home visit and indeed conducts the tour. Family members and caregivers may be present, but the therapist makes clear that s/he wants the patient to conduct the tour and do the talking. This contributes to patient activity and is likely to be an event that promotes the patient's emotional engagement with the clinician.

Typical clinic-based therapists have never had occasion to visit a patient's home. Consequently, the patient has never had an opportunity to show his/her strategies for dealing with disabilities or artifacts of life before the injury. Importantly, the patient often reveals unexpected strengths during a home visit. The therapist should be able to gain significant insight into the patient and the patient's physical, cognitive, and social environment during this visit.

An important part of the visit is the exploration of one or more candidate activities for therapy goals, an exploration that involves examining relevant artifacts. The current system for performing an activity is observed, and subtasks that need caregiver support are studied. If there is computer software that is being used, that will be observed and noted, especially for failures. At the end of the visit, the therapist will have decided on the initial intervention and will confirm the choice with the patient.

7.5   THE PATIENT DESIGNS THE USER INTERFACE AND SEES HIS/HER IDEAS IN THE SOFTWARE

The patient plays a major role in the design of the UI, which proceeds under the patient-centered design model, in which the patient's application functionality evolves and is driven by both the treatment plan and patient suggestions. Clinicians and technical staff make clear to the patient that the patient's judgments and ideas are valued, and the iterative process of UI design strongly reinforces that. The iterative process continues after the software is delivered and in fact proceeds throughout treatment. Patients have talked about their positive emotional response to having people associated with their treatment listen to them and take their advice. The UI is detail dominated, with few guidelines to inform CAT design at that level of granularity. Fortunately, patients, even those with severe and profound disabilities, have the ability to guide the design of the UI.

The goal of the UI design effort is primarily to reduce learning time, but it also aims at optimizing other performance measures shown in Figure 4.2. An important technique is to use the patient's words to frame text. This increases the closeness between the patient's mental model and the system's model, so that the patient's own words and concepts are used for the text of commands and instructions. The patient is given the time to redesign all aspects of the UI, including the size, shape, and location of various objects; the colors of objects; and system responses to user actions. With the the Institute's UI customizer, the modifications can be done in an interactive design session. The patient suggests changes in the interactive part of the usability testing session, and they appear on the screen. A few cycles of testing and redesign continue until a good-fit design is achieved. Additional changes can always be made.

In this approach, the software has an emotional impact on the patient because s/he sees his/her ideas incorporated into his/her own computer system and because this is a system that the patient can use effectively in performing activities important to him/her for the first time since the injury.

By this point, the patient has been trained to be active and will freely give opinions and suggestions in therapy and design sessions as s/he works toward achieving success.

Psychologist Dr. Brian Richards of Baycrest in Toronto and computer scientists Dr. Ron Baecker and Michael Wu, in collaboration with brain health specialists at Baycrest, have been successful in applying PD to individuals with senior amnesia (Wu et al., 2004, 2010). Some patients have become not only active but also engaged in the use of cognitive rehabilitation software (Michael Wu, Personal Communication August 2012).

7.6 DURING SESSIONS, THE PATIENT THINKS, TALKS, AND TYPES, AND THE THERAPIST THINKS AND TALKS

In conventional cognitive rehab, the patient comes to the clinic and is given an exercise to do with the therapist. The patient doesn't need to do any preparation, and there is rarely homework after the therapy session.

With the CPT approach, the patient is at home, sitting at a desk during the therapy session. The patient is an active user, doing the typing on forms and documents on the computer. Most activities, even those outside the home, involve some planning, organizing, and preparation that can be done in the home. The therapist doesn't type but typically asks the patient questions that help structure the tasks at hand.

The therapy session will end with the patient knowing what to do next. Early in therapy the therapist gives the patient assignments that the patient takes down in writing. As therapy continues, the patient has become sufficiently active to know what to do between therapy sessions.

7.7   THE PATIENT WORKS ON SHORT-TERM-GOAL ACTIVITIES BETWEEN THERAPY SESSIONS

The work between therapy sessions is part of the modality's design and is important. The structure of CPT therapy helps make the patient an active user, but both the relevance of therapy activities and the successes therapy brings help make the patient an emotionally engaged user too. Because the patient is at home, working at a desk on his/her computer, and because of the assignment, the patient is able to pick up where things left off. If there is a question in the patient's mind, there is a written assignment designed to move the activity forward.

As the patient progresses, the patient will work not only on the homework assignment but also on other activities that seem to be amenable to the tools and capabilities at hand. These are likely to be patient priority activities tied to some upcoming event and are likely to be part of STGs in the treatment plan. When the patient works on tasks and activities not directly tied to a recent therapy session, this is significant progress. The active user is sufficiently active and perhaps engaged enough to apply skills developed during therapy to other activities.

7.8   THE PATIENT GENERALIZES THE USE OF THERAPY TOOLS

Generalization has a special meaning in cognitive rehabilitation, because patients tend to have difficulty applying general principles to concrete situations. Generalization means that the individual can apply a tool, technique, or principle in ways not specifically taught.

Computer software presents an additional layer of complexity. The individual needs to learn not only commands and their syntax but also how to map the task onto the software. For example, typing a list requires a different mapping from typing a letter and from typing a set of instructions.

Individuals can show generalization in three ways with CAT. All of them reveal active-user levels of patient involvement. The first way is by expanding the scope of an activity. Individuals with cognitive deficits will often reduce the scope of an activity they used to perform by eliminating some of the subtasks that used to be part of the activity. This reduces the complexity of the activity. In therapy, the patient might be taught how to use a technological scaffold to support the performance of the basic activity. If the individual then figures out how to apply the scaffold to the expanded activity, that would be an example of generalization.

A second form of generalization is applying a technological scaffold to a different type of activity. If therapy involved writing a letter, generalization would involve another type of writing activity such as creating a shopping list (and note that different kinds of shopping will involve mapping the activity in different ways).

A third kind of generalization involves asking for additional features that would enable the individual to expand the scope of an activity or would enable the individual to perform a different activity with the new set of tools. This is a major demonstration of generalization. The individual not only shows a knowledge of the tools but also is able to see how the tools can partially assist in performing a new activity, and what additional kinds of features would enable the performance of the activity. Seeing the need for additional features and asking for them is considered a major achievement by the patient in the use of this modality. The patient will have been involved in many other instances of being an active user before this point.

7.9   THE PATIENT IS ABLE TO SEE FREQUENT TANGIBLE PROGRESS IN ACTIVITIES S/HE CAN PERFORM

The technology and therapy combine, in most patients, to produce ongoing tangible progress in the priority activities they can perform. The patient intake process and preliminary stages of software development train the patient to be active. The initial intervention is designed to achieve rapid success. Success breeds success. And the generalization process adds to the repertoire of subtasks that the patient can perform. Additional priority activities come into therapy as the patient progresses. The patient also experiences frustration during the course of therapy, but the therapist deals with it. And when this happens, the therapist is able to point to the new baseline of activity that the patient has achieved. Using this modality, all of the patients who make clinical progress become active users.

7.10   THE THERAPIST ASKS FOR SOFTWARE FEATURES AND APPLICATIONS THAT CAN ADVANCE THE PATIENT'S PROGRESS

Therapists who use CPT technology are active users too—conducting therapy is an active process.[7] Therapists will decide on the activities that will become part of therapy sessions and need to anticipate software features to be added to support the patient. From time to time, the therapist will see that neither conventional tools nor present CPT tools have the features needed to help the patient progress and will see that a new feature is needed for the cognitive prosthetic software. The therapist will provide the information needed to specify a user requirement, which will be sent to an application developer.

It is very important for therapists to be able to have features designed in this way because it pushes the cognitive rehabilitation envelope in that a tool will be available to solve a clinical problem that could not have been addressed before. Many CPT tools have been developed this

---

7 Some therapists are not well suited to this modality of therapy. Therapists need to plan and prepare in advance, because therapy goals and sessions relate to patient priority activities and must have continuity. Therapists also need to be spontaneous and be prepared to travel down a path the patient suggests.

way. The CPT software architecture is designed for enhancement, which is now easier with agile software development tools.

Therapists are active users and can become engaged users by experiencing the successes of their patients. This is all the easier because of the slow and limited progress so many brain injury patients achieve with currently prevalent therapy techniques. Therapists also respond with engagement to their ability to add enhancements and applications to CPT.

7.11   THE IMPORTANCE OF THE ACTIVE USER AND THE ENGAGED USER IN PATIENT REHABILITATION OUTCOMES

The significance of the active-user and engaged-user concepts is their relationship to two kinds of cognitive rehab outcomes: cognitive abilities gains and cognitive functioning gains.

Cognitive abilities gains represent a partial cure, an observed increase in cognitive functioning that is unaided by technology tools. These are increases in cognitive abilities that appear over time as a result of activity repetition in this modality of therapy. The combination of individual engagement, salience of the target activity, and repetition are elements that have been linked to brain plasticity (Cramer et al., 2011 and Walpaw, 2012). Developing an active user and then an engaged user is seen as contributing to this cognitive abilities gain. The combination of rehabilitation therapy and technology that is adapted to support treatment goals stimulates the use of cognitive abilities. It is important, as we have said over and over again, that these goals address priority activities inside the patient's planning horizon, activities that have high salience to the patient. It is likely that some of these abilities had declined because of nonuse, the cognitive equivalent of muscle atrophy from nonuse. Another possibility is that cognitive abilities have declined because of damaged cognitive elements in the brain.

Cognitive functioning gains result from augmenting existing cognitive abilities with well-designed technology, especially a patient-designed UI, and therapist support.

7.12   SUMMARY

This chapter has discussed ways in which patients become active users and engaged users through cognitive prosthetic telerehabilitation. The process of therapy is structured to have the patient become increasingly active cognitively, with frequent rewards that have the potential to increase activity and promote engagement. The therapist, using clinical skills, structures situations in which the patient can begin performing subtasks and tasks, from conducting the home tour to identifying priority activities to designing and redesigning the UI to focusing and typing during sessions to working between sessions to generalizing tool use.

This has generally come at a time in the patients' lives, and the course of their disease, when treatment has failed, and the realities of the caregiver process don't give them the time they need to perform some activities that now require more time. The active user and especially the engaged user with cognitive disabilities has the prospect of increasing cognitive abilities, perhaps through natural recovery and perhaps through a neuroscience mechanism. This stems from and is fostered by their activity and their engagement.

CHAPTER 8

Patient Case Studies in the Use of Cognitive Assistive Technology: Successes and Failures

This chapter adds to the case studies presented in earlier chapters. Combined, these case studies constitute data on patient outcomes, and in many cases, how they were achieved. The case studies in this chapter provide different "slices" of data. One provides log data of patient performance with timestamps down to the second. Another case study presents the beginning of treatment, some description of treatment, and the successful completion of treatment. The set of case studies provide the reader with a breadth of information about patient progress and therapy provided. While the cases come from different eras of technology and therapy, they share the key factors. These are the following:

- The patient was provided with a computer to use in the home between therapy sessions

- The computer had cognitive prosthetic software, and the user interfaces were personalized by the patient

- The computer was used for therapy sessions

- Patient priority activities were used as part of therapy

- There were telerehabilitation sessions into the patient's home

While there were differences between patients, these are relatively minor and the case studies can be considered to be replications of each other.

8.1 CASE STUDIES

8.1.1 MELISSA

Melissa was a 34-year old woman with a high-power fashion career who suffered four strokes before her condition was recognized and easily controlled. These left her with profound physical and cognitive disabilities. Her inpatient and outpatient rehabilitation history was extensive in physical and cognitive areas; at the time of her evaluation, no further cognitive rehabilitation was indicated. Cognitive disabilities included executive dysfunction, poor information retrieval, slowed information processing, impaired visual and verbal memory for new information, and impaired reading and word retrieval. She couldn't use one hand at all, and only had limited use of motion and use of two fingers on the dominant hand. She was dependent on others for dressing, feeding, toileting, and bathing. She could write with difficulty, and could walk short distances in her home. Her reading ability was limited. She was able to recognize 50% of the words in double spaced text. She was moved back with her parents who were devastated by her condition. Her life was an open book and she wished to be able to write and express herself without others having access to it. Her strengths were her ability to verbalize her thoughts, strong motivation and tenacity to improve her level of functioning, broad interests, and enjoyed physical exercise.

A basic word processor was designed with password protection for logon and document. The original word processor's commands were Print, Save, Retrieve, and Exit, all implemented by function keys. Three punctuation marks were implemented as remapped lower-case keys, so that she could use one finger to type. Documents needed to be double-spaced.

The results were substantial to the point of being several anomalies. In the first two days there were measureable spelling improvements, and in the first two weeks there were measureable grammar improvements. Also, within the first few days, she increased her stamina and perseverance for writing and reading—two cognitively demanding tasks—and her effective working time increased to three to five hours.

There were broad and unexpected improvements in her cognitive and physical functioning. After three months, her visual scanning improved so that she was able to use single-spacing in her documents. Additionally, she had become able to read 100 book pages in three hours, impressive for the reading speed and the stamina.

Physically, she made major advances in the use of her "good" hand. She was able to toilet independently at night, and during the day with the right clothes. She was able to bathe herself and to shampoo her hair. She was able to dress herself with the right clothes. She was able to eat finger foods herself. She had regained a privacy, control, and independence that she had not experienced since her illness. Her advances were attributed to brain plasticity by a leading researcher in that field (Merzenich, 1993).

After two years it was decided to mainstream her to Microsoft Word, and to introduce her to the World Wide Web, and she did these successfully.

8.1.2 HUGH

Hugh was a 61-year-old man, bicycling down a hill at 35 mph, wearing a helmet (which saved his life), when he hit a pothole, and was thrown 30 feet. He was in a coma for a week, spent several weeks in an outpatient "day-hospital" cognitive rehab program (six hours a day, five days a week). He did not have a good fit with that program, and in the eighth week his insurance company referred him to the Institute. He was a senior VP of a multi-national bank. In the initial interview and evaluation, he was in a cognitive fog, aware of the people around him, but needing substantial support. He was able to carry on a conversation, but had difficulty answering many questions about his life, and was slumped in an armchair. When the conversation turned to his work, he sat up, his body language changed and his speech changed. He was able to answer questions in detail, with confidence in his voice, and with changing intonation in his voice. This was a different person.

The day hospital program had him doing exercises that were difficult for him. He recounted an exercise where his three-person group was to order pizza for the people in the program. They were able to locate the address of a pizza parlor, and its phone number. They asked how many people needed to be fed. From that point on, they received negative feedback. Did they know how many people wanted what kind of pizza? Did they know if the pizza parlor delivered? What hours was it open? A woman on welfare assumed the role of group leader. While Hugh was exasperated and angry at what he saw as trivial questions, this woman stayed calm. He admired her for her ability to deal with them. He didn't feel that the therapists were helping to work through a problem. Rather, he felt assaulted, and left the program. He next went to a local neuropsychologist. Among other things, she gave him an assignment to select any article in *The Wall Street Journal*, read it, and list five important points in the article. He read *The Wall Street Journal* every day on the train to work. This was background information for him; he never consciously noted important points though he might see articles to chat with others about. He felt he would be graded by the article he selected (no), and by the points he selected (only to the extent of their importance).

The Institute was next, and our approach is patient-centered. He and his wife had many friends and colleagues who wanted to visit. Hugh wanted an application which would show who was scheduled to visit and when, and who had already visited. This application he could view, but didn't do any data entry; family members would do the data entry. Hugh was pleased with this, and would consult it frequently. He also expressed an interest in seeing his friends and colleagues at work. It was arranged for him to go into work a couple of hours a day, a couple of days a week. One day he happened to be there for a meeting, and sat in. He didn't speak, but commented to his therapist that he had no trouble following what was said. When it came time to think about reading material, we had the benefit of learning from others' mistakes. It was decided to have him read the weekly briefing book, which had the portfolio of large loans and a current narrative. His primary therapist, a Ph.D. psychologist, had never taken a business course, and couldn't determine if he was correct in his interpretation of the numbers and the text. This became an advantage. They would go through the book together for a couple of hours at a time, and Hugh would explain the contents to the psychologist. He was reading and explaining, but she didn't need to comment or answer questions. The third week as he was reading and explaining, he suggested that it looked like the bank was fixing its balance sheet in order to sell the bank.

Reading *The Wall Street Journal* wasn't that far removed from the weekly briefing book from the psychologist's perspective, but it was far removed from Hugh's perspective. And this was instructive.

This was also a radical departure. By convention, the therapist needed to be able to judge the accuracy of the patient in reading comprehension. In this situation though, the material that was most appropriate for the patient was material that she couldn't evaluate. Hugh started attending the weekly meetings where this book was reviewed and the account discussed.

Cognitive rehab follows an education model where there's a progression from easy to hard, from a therapist's perspective. In this case, the briefing book was easy, and other materials were hard.

As it turned out, Hugh was able to deal with abstractions, calculations, other's motivation, but he couldn't go into a convenience store and buy a bottle of milk and loaf of bread. He became overwhelmed by its complexity. Again, what was objectively hard was easy for him, and conversely, what was objectively easy was hard for him.

He increased his time at the bank, and was then allowed to take on some responsibilities. He was given a difficult project, and was able to do well with it, according to his manager. Not long after the completion of that project, it was announced that the bank would be sold.

The goal of cognitive rehabilitation is to restore function to the individual. Typically this is done in stair-step method, with mastery of easier material leading to the introduction of more difficult material. Return to work is considered the final step, where all of the reacquired skills can be applied to this most difficult of tasks.

Our approach was "out of sequence." Return to workplace had a strong and valuable social component, along with *exposure* to the cognitive content of work. Additionally, for Hugh, what was hard was easy, and what was easy was both hard, and largely irrelevant. In the end—actually long before the end—he was able to show and nurture his cognitive capabilities.

8.1.3 JANE

One afternoon in 1993, a Bell Labs HCI group called. One of their scientists saw our poster presentation at the recent RESNA conference and thought I might be able to help a 31-year-old scientist who had a stroke 15 months before. It looked like "Jane's" brief career had come to an end—the window for recovery from stroke was 12 months—and as HCI scientists, the Bell Labs group believed our unique approach of HCI coupled with therapy could. Her main problem was difficulty understanding what she read coupled with difficulty in language expression; she was aphasic. I politely thanked them for their interest, but we only worked with traumatic brain injury patients, and only nearby, and had never encountered aphasia. Her colleague made an emotional and flattering case, along with the promise of communications and technical support.

The decision to explore working with Jane opened the door to what is known as the acquired brain injury field (ABI) and the distance therapy (telerehabilitation) field. Two hours away was at the outer bounds of commuting distance. Our clinicians were familiar with stroke, which was treated by the same physicians and therapists who treat traumatic brain injury. Stroke was new to the computer scientists. A state-of-the-art Kurzweil reader was made available to us, and could do text-to-speech.

Jane was an avid sports fan and was unable to read the sports pages of newspapers or Sports Illustrated. At one point, we observed her flipping pages in a manual for a handheld-computer sent to her by Bell Labs; handhelds were uncommon in 1993. Computer manuals are generally considered more difficult than sports pages, and she was asked what she was doing. She replied that she was reading the manual to learn how to use the machine, and indicated that she had good comprehension.

It seemed clear that this was an instance of "what was hard was easy, and what was easy was hard." It was decided to give her an article in a top journal to read. A methodological article was selected that was of interest to researchers in her section. They were asked to supply questions about every two paragraphs. The article was digitized, and placed in a font, size, and color that Jane selected. However, the Kurzweil reader was just that, it could read the contents of a file, and locked use of that file, preventing Jane from typing her answer to questions. We had had previous experience trying to modify a device of that era. Our project was to help Jane, so we turned to solutions that could be used with a DOS operating system and a word processor. A version of an AT&T text-to-speech unit was provided, and some of the phonemes (phonetic sounds) were adjusted.

An application was assembled that had sufficient word processor functionality, could do text-to-speech each sentence, as well as highlight a sentence, and advance or repeat reading a sentence.[8] She was able to understand the article, and to provide adequate answers to questions her colleagues had prepared to test her comprehension. Here was an even more extreme example of "what was hard was easy, and what was easy was hard."

With this success, it seemed likely that she could be given a research task. When she was offered the opportunity, she accepted with gratitude. A project was selected that had no deadline pressure. And she was able to work on it, mostly telecommuting but also returning to the office.

8.1.4   DR. K

Dr. K was an emergency medicine physician in his mid 50s. Nursing staff noticed that he was having difficulty managing and treating his patients, and especially heart attack patients. He was also taking dinner two and three times a shift. He was surprised that the cafeteria kept on serving the same menu. The hospital responded quickly and appropriately. Dr. K was evaluated, and diagnosed with profound deficits of immediate and short-term memory, but his long-term memory was intact.

---

8 The Kurtzweil unit provided her was unable to read from "open" text files, which meant that she could not use it if she were going to type her responses to the questions that were embedded into the text.

One of the results of his neuropsychological testing was that he remembered no information after 30 minutes, quite devastating for acquiring new knowledge. However, the rehab hospital (not where he had worked) was a 40-minute drive from his home, through much traffic and other distractions. Yet he had no problem remembering that he was going to rehabilitation. This was an anomaly. Somehow, this portion of his memory was working.

His therapist was involved with the Institute, and thought that cognitive prosthetics software would be appropriate, especially for his schedule. In consultation with Dr. K, it was decided to link the appointment and schedule application with his alphanumeric pager, so that he could send himself messages. He and his therapist found this a successful approach, and with this, he was able to track his appointments and commitments. He was also able to reduce his cognitive load, and as a result, he seemed to be having more control over his life.

Emergency Medicine is an intense practice. However there are other areas of medicine that were very routine. This was pursued under a Physicians Health Program of the state medical society, which enables impaired physicians to practice under strict supervision of another physician. Protocols were developed for him that included physical examination, differential diagnosis, treatment, and documenting the case in the chart. He began working very part time, and continued adding hours successfully until he was working full time. Due to the nature of his impairment, he would always be working under supervision.

In summary, this case illustrated several points. The alphanumeric pager was a valuable tool for reminding a patient of an activity or event. The clinical cognitive testing process may indicate profound deficits, but the individual may still be able to perform effectively in the contexts of his own life. Finally, individuals may be able to function adequately when their cognitive load is in balance with their cognitive resources.

8.1.5 DICK

Dick was in his mid 50s and had a stroke. This kind of stroke often leaves the patient with a major disruption of episodal memory, memory related to the individual's activities. For 15 months he had received brain injury rehabilitation at two leading facilities, and there had been little change over the past several months. In his case, he could remember almost nothing of what he had done, or what he was supposed to do, but his language, personality, and other mental capabilities were intact. He was very pleasant in conversation and retained a sense of humor. He lived in an exurban home, and had a vacation home as well. Both he and his wife had enjoyed active outdoors activities.

When he entered the Institute's program, he could be left at home alone for very brief periods of time safely, but he would not be able to accomplish much, or even prepare simple breakfasts or lunches. They were both active in their church, and he was able to do tasks there with supervision, and enjoyed greeting people for services and events. His wife, with experience in dealing with cognitive disabilities, had tried to provide him with organization strategies using calendars and to-do lists, but they would get lost.

As for goals, his wife hoped that he could become self-sufficient during the day, and that she could continue her activities with her friends. They would also go to their vacation home. The initial goal was for Dick to be able to perform some activities self-sufficiently for several hours a day, with no direct supervision. He was handy around the house, and small projects could be organized. The Initial Intervention would be for Dick to be able to follow a simple schedule and perform some activities he enjoyed each day.

The prosthetic application began with a calendar for the day, and a few activities for him to do alone, and he would have short therapy sessions several times a day. He would carry his one-page schedule with him, make notes on it, and could print it out again when it was lost. He would also update it during the day, showing what had been done. Most of the initial projects were in the house or yard. His wife helped to suggest activities and projects, and observed him as well.

He responded well to the initial intervention. Also during this time, he would be dropped off at the office so that he could chat with former colleagues. He would not be returning to work, but the question arose of his ability to work on a technical project. A coworker suggested an actuarial problem that he would have been able to do, and prepared a copy—a complex Excel spreadsheet—for him. Dick worked on this with his therapist, using the same approach as with Hugh's weekly report. Dick worked on the spreadsheet, talking to the therapist. Because of the state of his episodal memory, she suggested that at the end of each session, he prepare a message to himself describing what had been learned, and what the next steps were. This was called the "Back to the Future" strategy, after the movie series. This exercise was a success. Dick discovered that there was some missing data—which unknown to him, his colleague had removed. When Dick wondered about the data, it was restored, and Dick went on to solve the problem.

There were two anomalies here. One was the length of time—one to two hours—Dick was able to spend on this project and still have the information in his working memory. Working memory is expected to hold simple information for half an hour or so. The contents of this spreadsheet were complex. However, having the open spreadsheet supported his memory during this activity. The second was his ability to problem-solve this complex problem, including the ability to determine that there was missing data.

Over a number of months, Dick's episodal memory continued to improve, and he continued to use the schedule. His self-sufficiency increased. His schedule served as a diary of things he had done. He was able to drive short distances with little traffic. He made more gains, and was able to drive into the city, through traffic. And he had become self-sufficient for his daily activities.

The other anomaly was Dick's recovery of function. There was some recovery of episodal memory that went beyond just relying on his calendar. He was active in its use, and his positive outlook on life may have helped him become and maintain engagement. Additionally, he was able to observe and experience continuing success. The therapy also was carried out in his own settings, and with his specific activities. He had superior intelligence and a supportive family. Nonetheless, this result remains an anomaly.

8.1.6 LESLIE

This case study focuses on two issues, deficits that emerge after discharge from therapy, and the therapist's use of two sets of tools, 1) data generated by patient software, and 2) therapist's use of the customization software that was able to personalize the patient's software modules in important ways.

Leslie had achieved the status of being "high functioning," testing as "normal" after cognitive rehabilitation outpatient therapy. She was independent in cognitive household and social activities. Then she discovered that she was having difficulties with tasks and activities at home. This is the syndrome where the patient succeeds in the clinic, but that success diminishes in the home. That is because there are aspects of performing an activity at home that were not considered in the clinic, as clinic-based therapists don't observe the patient in the home. Specifically, she wanted to write some letters.

Leslie and the therapist decided that letter writing would be an appropriate place to begin as a therapy goal. A four-function word processor was customized, tested, and re-designed for the Initial Intervention. The therapist and patient worked together on a letter, and the session ended with the expectation that Leslie would finish it over the weekend.

The therapist examined the usage log and saw that Leslie had spent two-and-a-half hours over two days working on the letter, and that it had been printed three times (Figure 8.1). During the next therapy session, Leslie explained that she wanted to make the letter "look right." That turned out to involve both changing the formatting with spaces and tabs, and checking the spelling.

```
5/30    8:46:00 PM     Select New
5/30    8:46:00 PM     Opened New Document
5/30    10:58:26 PM    Select Save
5/30    10:58:52 PM    Saved As - c:\My Documents\Dot and Joe letter.rtf
5/30    11:16:32 PM    Select Save
5/30    11:16:38 PM    Closed - c:\My Documents\Dot and Joe letter.rtf

5/31    1:45:36 PM     Select Open
5/31    1:45:37 PM     Opened - c:\My Documents\Dot  and Joe letter.rtf
5/31    2:04:31 PM     Select Print
5/31    2:04:31 PM     Printed –
5/31    2:08:55 PM     Select Print
5/31    2:08:55 PM     Printed –
5/31    2:22:03 PM     Select Print
5/31    2:22:04 PM     Printed –
```

Figure 8.1: Data showing two-and-a-half hours over two days.

The therapist saw a template as a means of addressing the formatting problem. This would require adding a **Save As** function. As for spelling, the therapist could add a **Spell-Check** feature. She discussed options with Leslie who agreed that a template would be a good approach, and appreciated having a spellchecker. They developed the template, and developed the UI for the **Save As** feature, and the spellchecker.

Cognitive prosthetic software is designed to be readily changed during the UI testing and redesign process, and when new features are added. The Customizer is the application that has a common interface that can personalize all of the patient software modules. The Customizer is designed to be a therapist's tool, either directly or in "chauffeur-driven" mode, where a tech support person applies to keystrokes to implement the design the therapist specifies (Cole et al., 2000).

She then trained Leslie in the two features. Leslie was given an assignment to write another letter using the new features. The next day the therapist examined the log. This letter had taken 20 minutes in one session to write, spell-check, print, and save (Figure 8.2).

Later, the patient was moved to MS Word, and then to Excel.

6/6	7:51:06 PM	Select Open
6/6	7:51:17 PM	Opened - c:\My Documents\Letter format.rtf
6/6	8:08:25 PM	Selected Speller
6/6	8:08:30 PM	Misspelled - NJ
6/6	8:08:31 PM	Skipped
6/6	8:08:41 PM	Misspelled – Raod
6/6	8:08:41 PM	Replaced - Road
6/6	8:08:44 PM	Misspelled - parent-educator
6/6	8:09:00 PM	Skipped
6/6	8:09:15 PM	Select Print
6/6	8:09:16 PM	Printed - c:\My Documents\Letter format.rtf
6/6	8:10:08 PM	Select Print
6/6	8:10:08 PM	Printed - c:\My Documents\Letter format.rtf
6/6	8:11:04 PM	Select SaveAs
6/6/	8:11:30 PM	Saved As - c:\My Documents\Jean.rtf

Figure 8.2: Patient log data showing 20 minutes.

There were two important outcomes here. The first was the ability of the patient to perform an activity in the home setting that she hadn't been able to do. She re-learned how to use computer software that most people in our society use.

The second outcome relates to the therapist, and her ability to use two therapist tools. The data both answers questions and raises questions. The therapist couldn't tell what the problem was with Leslie's initial effort at writing a letter. For that answer, she needed to talk with Leslie. The other tool was the Customizer. The Customizer both allows for the personalization of the UI and is the means of activating functionality in an application. This case study shows how the therapist was able to identify the needed functionality, and then to work with Leslie in designing an easy to use UI. She was able to use both sets of tools in a matter of minutes.

8.1.7 LAURA

Laura was a woman in her thirties who had a TBI from an automobile accident while drunk, and had several DUI convictions. She dropped out of high school and was a single mother of two teenage children. She was admittedly a poor homemaker. Part of her cognitive rehabilitation involved having her wash dishes in the sink and sort the clean laundry for her two children and herself. She hadn't been able to achieve the therapist's expectations in the homemaking tasks, and this prevented her from progressing in rehab, which would include return to work. This should have been a substantial motivator for her. Her rehabilitation nurse, talking to her employer, learned that Laura had a reputation for a strong work ethic and had been a good worker. Additionally, her employer's plant management, located in a rural area, had a strong sense of coming to the aid of people in need.

Laura was a production coordinator, and worked on the factory floor. Her job was to gather about a dozen materials for each order, and then schedule the run before a due-date. This required substantial executive functioning skills, which include, planning, organizing, and follow-up, as well as good communication skills. In addition, her desk was on the factory floor, which had the usual noise level of a fabrication plant. Her work tasks were far more difficult than her homemaker assignments.

She had been away on medical leave for nine months when she returned to the plant, to begin working a couple of hours a day, a couple of days a week. On that first day, she sat down and began going through the files and the schedule. She would get up and consult with individuals who needed to prepare a component for the job. Her performance that first day was remarkable, and after several weeks had worked her way up to full-time. However, her homemaking still remained problematic, and her children helped. Here again, what was hard was easy, and what was easy was hard.

8.2   FAILURES

There have been patients who have failed to have gains from this modality of therapy, and it is worth noting some of them.

8.2.1 LINDA

Linda had been in an automobile accident almost a decade earlier. She had cognitive strengths and a supportive family. However one of her deficits was her inability to recognize the rooms of her small home, meaning, among other things, she could not easily find the computer. This case was also hampered by having poor telecommunications in her rural home in the early 1990s. It was also hampered by having limited technology and limited clinical experience. However, it is quite possible that her pattern of deficits, which wasn't amenable to conventional therapy, weren't amenable to CPT either.

8.2.2   MATT

Matt had been in college when he had a TBI. After cognitive rehabilitation in his community, he returned to college. The Institute was contacted to provide cognitive rehabilitation and academic support in his dorm room 50 miles away. He appeared to have the cognitive ability to do college work. However, priorities were social, and he enjoyed being one of the gang. He understood that he was on track to flunk out, but he enjoyed the feeling of doing "normal" things with his friends. He seemed to need psychological services far more than cognitive prosthetics, so we discontinued services.

8.2.3   JACK

Jack had recovered medically from brain cancer, treated with surgery, chemotherapy, and radiation. Brain cancers are diffuse and his cancer was treated with diffuse radiation to the brain. His executive dysfunction was broad, and strategies were insufficient to make significant progress.

8.2.4   CLAUDIA

On referral, Claudia was described as having a number of cognitive disabilities, coupled with fatigue that made clinic-based therapy unsuccessful. She presented a number of functional problems, and suggested software functionality that she thought might help. Claudia seemed to have excellent insight into her problems, and a number of applications were designed and launched. However, she seemed to make little progress in resuming activities in her life. As time went on, more applications were developed. She had an ability to identify problems needing remediation that didn't match her ability to resume productive activities in her life. There were early indications that suggested she was emotionally invested in her symptoms, and there was always another condition or problem that could be identified.

8.3   SUMMARY

Case studies in this chapter span the range of outcomes, from Melissa whose major recovery was attributed to brain plasticity to those who showed no recovery. These case studies also differed in the kind of data presented. Leslie's case presented system log data providing a window into process, down to the second, plus the therapist's intervention with cause and effect from software functionality. Dick's case study showed a remarkable recovery of function and independence. That data mainly consisted of beginning and end of therapy, without possibility of spontaneous recovery. Anomalies have been present in several of these case studies, or rephrasing, the use of cognitive assistive technology has been able to produce positive results not explained by current theory.

Some failures were presented. All of our patients had failed at other programs. In that context, even small successes attributed to technology would be significant; if that glass became one-eighth full, that would be significant. The four individuals in the failures group all gave early indications of failure. One wonders whether more clinical experience or new generations of technology would have changed the outcome. It will for some.

CHAPTER 9

Conclusions, Factors Influencing Outcomes, Anomalies, and Opportunities

This chapter provides a summary of the major concepts and outcomes from the earlier chapters. This book has explored cognitive assistive technology's power as a therapy tool in cognitive disabilities from brain injury. Over a quarter century, the Institute for Cognitive Prosthetics has applied Human Computer Interaction models and methodologies in evolving technologies that increase the level of functioning of individuals with cognitive disabilities from brain injury. Over that period of time, there have been vast increases in information and communication technologies. However, there have been common elements to each individual's treatment. Each was given a computer for the home. The computer had their personalized cognitive prosthetic software. Each was working on an activity of interest to them. Each individual's activity was one that they were unable to achieve themselves without assistance from others. Each was heavily involved in the design of their user interface to produce a good fit between the patient and the user interface. Individuals generally received treatment for many months, and in some cases several years.

The case studies show strong results for a patient population with chronic cognitive disabilities, strong enough to warrant more study especially in this area of critical need and problematic medical efficacy. The Institute's goal has been the development of technology-based techniques that have substantial ecological validity, where patients are able to increase their level of participation in everyday life. These results have been achieved for a significant percentage of patients rather consistently; there is a need for collaborators who can place patient progress in a cognitive and physiological context. A measurement problem has been to find clinical measurements that are at the same scale as the short-term treatment plan goal. A section of this chapter addresses the need for more rigor in exploring cognitive prosthetics telerehabilitation process and outcomes.

9.1   CONCLUSIONS

This section presents conclusions drawn from the case studies presented throughout the book, and suggests factors that may explain those outcomes. Significant anomalies in patient performance are found in a number of case studies, and are summarized.

9.1.1 THE USER INTERFACE IS THE MOST SENSITIVE DESIGN ELEMENT IN APPLICATIONS FOR INDIVIDUALS WITH COGNITIVE DISABILITIES FROM BRAIN INJURY

The user interface is the most sensitive and important design element in applications for individuals with cognitive disabilities from TBI. Furthermore, UI design is capable of providing a software bridge over some of those disabilities. This has implications for the field of cognitive disabilities from brain injury, and creates new opportunities for HCI designers and practitioners.

This conclusion is based on theory and on empirical findings. An early HCI model of interface performance (Card, Moran, and Newell, 1983) anticipates these results (Figure 4.3). This is a heuristic model relating user characteristics and system characteristics to UI performance. User characteristics include the vast range of cognitive dimensions related to computer use, and system characteristics include the UI design. Case studies above demonstrate that UI redesign guided by the patient is able to produce software with UIs appropriate for that user.

A question arises whether principles developed for DOS environments in the 1980s through mid 1990s are still valid today. Each UI presented in this book went through a personalization process to make it intuitive to the individual. That involved design/redesign cycles guided by the user until the design could be used effortlessly. It was not clear then what variables the user was evaluating in designing her/his UI. But the patient is the best "meter" of their own UIs. This is fortunate because the task of developing usable guidelines across the cognitive domains is Herculean. Since the mid-1990s, UIs could be designed interactively with the patient. This increased the speed of achieving an intuitive UI.

9.1.2 COGNITIVE ASSISTIVE TECHNOLOGY AS A VALUABLE THERAPY TOOL FOR BRAIN INJURY COGNITIVE REHABILITATION

Cognitive assistive technology has demonstrated an ability in a number of cases to apparently increase an individual's basic cognitive functioning; we have been able to replicate our findings. This has the potential to be a game-changer for CAT. As computer scientists, we note this observation, but don't have the theoretical or methodological tools to delve into the underlying mechanism. The tools are evolving in the neuroscience field.

Cognitive assistive technology is rarely portrayed as a tool for functional rehab therapy, based on everyday activities. Our early results clearly showed CAT's ability to be a tool for brain injury cognitive rehabilitation. This was clearly demonstrated in 1992 with the first NIH study described in the case studies of Suedell, Roy, and Sarah. All three subjects were able to increase their level of functioning with their cognitive prosthetic software within the first two weeks; each of these three had failed to achieve their goal after months of conventional therapy. Roy, in the first month of use, had an increase in time planning horizon that expanded from two weeks to eight weeks. This was an anomaly, since it was not merely technology support. Suedell showed substantial increased cognitive functioning during the study; gains in cognitive abilities over the next two years allowed her to live independently, six years after her TBI. Sarah was able to make gains in functioning, attributable to the technology. These patients made remarkable advances in their level of functioning, and this alone justified the involvement of therapists and therapy in CAT. It is clear that therapists and therapy add value to cognitive assistive technology, and CAT has added value to the therapy.

It is likely that there are other approaches to cognitive assistive technology that also can serve as therapy tools for brain injury rehabilitation. Replications are also a means of advancing knowledge in ICT fields. Replications can involve either the initial brain injury population or what is seen as a similar clinical population with cognitive disabilities.

9.1.3 PATIENT-CENTERED DESIGN AS AN IMPORTANT DESIGN METHODOLOGY WHEN CLINICAL ISSUES ARE A FACTOR IN THE DESIGN OF CAT

Patient-centered design is appropriate for scenarios in which clinical considerations are involved in the design of cognitive assistive technology. In particular, PCD is appropriate when CAT is being used as a therapy tool for brain injury cognitive rehabilitation. Cognitive assistive technology and Patient-Centered Design combine to enhance clinical outcomes and to increase technical outcomes as well.

PCD changes both the design process and the therapy process. The design process changes for three reasons. First, current (weekly) therapy session goals drive software functionality. Second, the functionality accumulates as new therapy goals are added. Third, the patient plays a major role in designing and redesigning the UI.

PCD focuses on the individual patient. In the clinical context, the patient has a medical problem, but also has abilities—some stemming from nuances of the condition—that are able to help in therapy. A focus on the individual patient can make abilities more prominent. From a developer's perspective, harnessing those abilities reduces the difficulty in designing a good and satisfactory solution. PCD is also able to help the designer explore the domain for a particular condition, and in the context of therapy, develop robust tools that are able to achieve success in a broader range of cases for the condition.

PCD changes the therapy as well. First and foremost, the patient does not need therapy sessions of training on a device; the technology is personalized and, typically in minutes, made intuitive to the individual. Therapy is aided by the patient's CAT, which is able to incorporate complex support for the patient who substantially exceeds the capabilities of conventional therapy techniques. Therapy goals use patient priority activities as the context for therapy. Goals can also be achieved faster by the patient's use of the prosthetic software between therapy sessions for their own activities. Therapy is aided by the patient's working with software s/he has designed and in which s/he takes substantial pride. This can deepen by adding additional features to the patient's prosthetic software, and the patient's contribution to that design.

Many of the patients exhibited an "if only" behavior, if only they were able to perform a specific subtask, they expected that they would be able to perform the desired activity self-sufficiently. "If only's" are also the individual's continuing attempt to overcome some of their disabilities and to regain greater independence. "If only's" can be viewed as a patient's continuing motivation, motivation available to be harnessed. PCD is excellent for providing those "if only's" by providing the limited functionality the individual desires, and using interactive and iterative design methods to help the user refine the enhancement.

Some work in the CAT community can be seen as compatible with the PCD model, such as work on aphasia by McGrenere et al. (2003), Moffatt and Davies (2004) and in senior demintias (Baecker, 2007; Wu et al 2004, Wu et al 2010, and Mihailids, 2011).

Each individual's CAT implementation is built from the ground up, one short-term therapy goal at a time. And each therapy goal is tied to a patient-centered activity, one that is a priority to the patient and one that will involve the patient in the very near future. Thus, the content of therapy sessions continues to be relevant because they are part of the patient's life.

PCD with its emphasis on clinical context can be adapted for cognitive disabilities associated with a specific disease and condition. There are differences across the diseases and conditions that cause cognitive disabilities and across the cognitive dimensions themselves, but there is every reason to believe that personalizing the UI has benefits for other cognitive disabilities.

9.1.4 THE ABILITY OF THE BRAIN INJURY PATIENT TO PLAY THE KEY ROLE IN DESIGNING HIS/HER UI, WHICH WILL BE INTUITIVE TO THAT USER

An important finding has been that cognitively impaired users are able to play a key role in designing their own UIs. Patient-Centered Design provides the opportunity for applying this finding. The finding is important on technical grounds because there are no adequate models to substantially inform UI design for individuals with cognitive disabilities from TBI. Our users tend to be precise in their instructions, and it is clear that something is being optimized, but exactly what is often elusive. This creates a UI that is intuitive to the user for their purposes, and as a result, requires virtually no training. In other words, the UI is highly efficient for the individual.

9.1.5 COGNITIVE PROSTHETICS TELEREHABILITATION IS A CAT AND A MODALITY OF COGNITIVE REHABILITATION THERAPY

Cognitive Prosthetics Telerehabilitation is the generic name for a modality of cognitive rehabilitation. It is an updated version of our Computer-Based Cognitive Prosthetics, defined in Cole (1999). Its success with brain injury rehabilitation is attributed to the synergy from the various elements. CPT is defined in Figure 6.1. The Institute's implementation of CPT has additional features. Key features of CPT are the following:

- Brain injury rehabilitation services become widely available, via broadband Internet into the patient's home.

- Cognitive prosthetic software is personalized for each patient using Patient-Centered Design.

- For the patient, the cognitive load of in-clinic cognitive rehab is eliminated, and the user experience is advantageous compared to in-clinic therapy.

- For the therapist, it becomes practical to superimpose therapy on top of the patient's actual activities, making therapeutic interventions have a better fit with the patient. The user experience for the therapist is also improved compared to in-clinic rehab.

- Therapy sessions are conversational, with patient and therapist at their respective work-stations.

- Most activities involve planning, organization, and preparation. These subtasks are well suited for therapy sessions in the home.

- The Institute's implementation of CPT features a number of therapist-friendly tools for monitoring patient activity, planning therapy, and reporting to third-party payers.

- CPT is sufficiently flexible that it can be implemented in ways that are supportive of differences across organizations.

Telerehabilitation isn't new, with most distance cognitive rehabilitation services delivered to a satellite clinic or prison. The home is a developing area (see Constantinescue et al., 2010). There are suites of software designed to provide broad functional support to the brain injury rehab patient (Kirsch et al., 2004). There are others who have customized software to the needs of a specific subject. Most therapists would describe their therapy as conversational with the patient. Many therapists would say that their therapy activities help their patients in planning, organization, and preparation. There are others who have specialized patient workstations, available from many venbdors, placed in local institutions rather than patient homes (Constantinescue et al., 2010). There are, however, few therapists who are able to base their therapy on a broad set of patient priority activities, and that can be easily explained by the difficulty of treating patient's actual activities from the confines of in-clinic therapy.

The case studies presented show a pattern of success with our earlier modality, continuing with our implementation of CPT. We propose that is the result of synergy from the elements that are part of CPT—that the components interact to produce an impact much greater than the sum of its parts—and produces the case study results. This is the fundamental difference between CPT and therapy based on some of its elements.

9.2   FACTORS INFLUENCING OUTCOMES

There are a number of factors which influence the results reported in the case studies.

9.2.1   THE THERAPIST AND TECHNOLOGY ARE BOTH ESSENTIAL PARTS OF THE CPT MODALITY

The most important factor is the marriage of therapist with cognitive prosthetics telerehabilitation technology in treating patients. This has been discussed above.

9.2.2   "PRESSING THE PATIENT'S BUTTONS"

Different patients are attracted to and mobilized by different stimuli. Color, music, cute cartoon icons, and animation have been able to dramatically attract the attention and focus of some patients, and have been able to "press their buttons" in a positive way. In UI design sessions, patients will often reveal the kind of UI elements (stimuli) that will enhance the effectiveness of the UI and the application.

9.2.3   THE ABILITY TO SUPPORT A BROAD RANGE OF PATIENT PRIORITY ACTIVITIES

For relevance to therapy, CAT needs to be able to support a broad range of activities in the patients' lives. TBI is one of many diseases that produce a broad array of cognitive deficits that affect a broad array of activities. Therapists need software that is able to serve most of the breadth of their user's needs. This kind of functionality can be supported by a cognitive prosthetics software suite, which includes a personalization module capable of designing user interfaces, and turning functionality on.

Patients like to economize on applications, expanding the features of one that they already know how to use. Often a particular activity can be supported by more than one kind of application. Patients, seeing the need for a feature, may request that feature become part of an application they are already very familiar with. An advantage of using a suite is that diverse functionality may be added to an application.

9.2.4   THE ACTIVE USER AND THE ENGAGED USER

From the first patient, the structure of our software design and therapy processes have encouraged patients to become active and engaged. Patients have appreciated the continuing integration into normal life that CPT provides. Making the patient active on therapy activities especially between therapy sessions has clinical and financial effects. The amount of time the patient spends on therapy-related activities increases when therapy activities and the individual's life activities are one and the same.

9.2.5   THERAPIST'S USE OF PATIENT WORK-PRODUCT DATA

Cognitive prosthetic software collects a substantial amount of usage data, with logging that is common in application development and testing. From the beginning of our activities, this usage data has had a substantial clinical role, and was responsible for identifying a significant anomaly in Gail's initial use of her initial application software.

Work-product data enables the therapist to look at patient performance between therapy sessions, and identify advances as well as potential problems. This is another avenue for understanding patient performance in the patient's settings. The therapist is able to explore possible problems, and then to respond to them, as in the case of Leslie, where corrective action can be taken.

This data can also serve as an early indicator of the likelihood of success. Alternatively, it can provide a warning of potential failure, as in the cases of Linda, Matt, and Jack. Claudia is an example of subtleties that are found in the complexities of TBI patients. CPT also has provided suggestions of malingering and instances of the presence of other relevant issues.

9.2.6   THE ABILITY TO SUPPORT THE PATIENT IN THE SETTINGS WHERE THEY PLAN THEIR ACTIVITIES

Without question, the ability of the therapist to work in the patient's setting is a powerful therapy tool. This enables the therapist to work on the patient's actual activities directly, and make therapy directly relevant to the patient's life. By selecting a patient priority activity, the therapist is able to make the therapy even more relevant to the patient. Therapy in the patient's home enables the patient to continue working on activities that were begun during the therapy, a seamless continuity.

9.2.7   LIKELY MECHANISMS

All of the the individuals who became the Institute's patients had failed at cognitive rehab in one way or another. It is quite likely that most made gains in previous rehab programs. And most patients made gains, some significant and some extraordinary. All patients received therapy involving highly customized technology plus therapy designed to take advantage of the technology; the first patient received only the technology. There are three mechanisms that may explain the gains. All of the mechanisms rely on the integrated combination of therapy and technology.

Cognitive Productivity Increases Due to Technology

Much of the software people use increases cognitive productivity, the ability to perform activities as a result of the technology alone. There is no apparent gain in cognitive abilities. The case of individuals with cognitive disabilities is considered similar in that increases in cognitive productivity are considered purely the result of the technology. However, the degree of cognitive productivity increase varies with the quality of the fit between the individual's cognitive deficits and abilities with the software.

Cognitive Load Reduction

A second mechanism is cognitive load reduction. The highly personalized technology that produces a better fit between the user and the technology reduces the cognitive load on the user, and with it stress. Additionally, the ability of the individual to do more activities in priority areas both adds to self-esteem and reduces the degree of reactive depression. All these combine to reduce the cognitive load, freeing up brain resources to apply to cognitive activities. As a result, the individual is able to increase cognitive productivity.

Cognitive Neuroscience

Cognitive neuroscience is changing the landscape of many areas of psychology (Raskin, 2011). Advances in neuroscience are responsible for removing the several-year boundary for possible recovery of cognitive functioning. Brain plasticity is now recognized as an important mechanism for cognitive recovery, although it is rarely seen in cognitive rehabilitation.

Wilson et al. (2001) and Robertson et al. (2002) have found that CAT interventions were in some cases able to produce changes in abilities that persisted after the use of CAT was discontinued. Evidence of cognitive abilities gain, especially where none is expected, is important and worthy of substantial research oriented toward discovering the conditions under which cognitive gains can be attained.

Merzenich, a leader in the brain plasticity field, associated Laura's cognitive and physical gains to brain plasticity (Michael Merzenich, Personal Communication)

Current theory suggests that brain plasticity responds to an environment where the individual is emotionally engaged, where the target activity is highly salient to the individual, and where there is repetition in the activity (Theodoros, 2011). CPT is able to produce those three behaviors.

Collaboration with neuroscience researchers may help determine the source of CPT patient outcomes that show evidence of cognitive recovery.

9.3   ANOMALIES IN PATIENT BEHAVIOR

Anomalies in patient behavior are different from what theory and experience anticipates. And they are worthy of further exploration and attempts at explaining why the anomaly occurred, and when it is positive. A technology that frequently produces anomalies—especially positive ones—is valuable.

9.3.1   THE SUBSTANTIAL ABILITIES OF INDIVIDUALS WITH PROFOUND AND SEVERE DISABILITIES

Before meeting Gail (Chapter 3), a packet of clinical evaluations and discharge summaries arrived. The evaluations systematically describe the individual's deficits and limitation, and briefly mention the patient's abilities. Part of the focus on the negative is that clinical reports are the basis for obtaining reimbursement for clinical services. Family members will also describe the caregiving tasks, with examples of how the individual failed, and the burden fell on caregivers. Also, they live with the difference between what the individual had been and what the individual is now.

Expectations had been set by those reports. However, in working with Gail, adaptations to Usability Testing provided the opportunity for her to demonstrate her substantial abilities in UI design/redesign, and were a counterpoint to clinical evaluations. This became a trend as more experience was gained from more TBI patients with severe and profound cognitive disabilities. Meeting with the individuals in the home typically provided the opportunity for them to demonstrate some of their abilities, as they described how they try to cope with some success to their disabilities, and what they are able to do without caregiver support.

Additionally, Gail and other patients were able to perform other activities that could be classified as anomalies of behavior in the presence of severe and profound cognitive deficits. Some of these involved using application functionality that hadn't been taught to them, or novel ways of using application's functionality.

9.3.2   THE LIMITATION OF "HIGH FUNCTIONING" INDIVIDUALS

"High Functioning" TBI patients are individuals whose recovery places them in the "normal functioning" range. It was surprising that "high functioning individuals typically required UI personalization to be effective in enabling the individual to use the functionality of the application. It was also notable that the interface modification typically enabled the individuals to overcome the cognitive issue quickly and dramatically, as in the case with Sarah.

9.3.3   ISLAND OF ABILITIES IN SEAS OF DEFICITS (IASODS)

Working closely with patients and observing them and their work revealed a pair of anomalies, this one and the next. Islands of abilities in seas of deficits on a cognitive dimension (IASODs, Figure 9.1) are cognitive skills and abilities that the individual has that are unexpected from clinical testing or from observing the individual. They occur within a single cognitive dimension, such as memory, language, problem solving, attention, and self-monitoring.

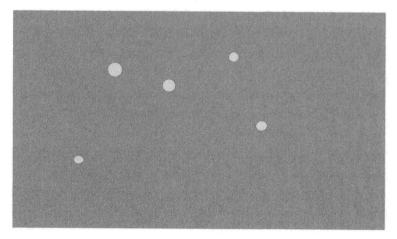

Figure 9.1: Islands of abilities in seas of deficits (IASOD).

IASODs are important because they provide toeholds in areas of weakness. In terms of computation, they provide focused abilities that can be leveraged and enhanced in designing prosthetic solutions or interventions. Computation has the ability to identify IASODs because of the difference in granularity found in clinical assessments versus granularity needed in application or UI redesign. IASODs have the potential of simplifying the amount of design that needs to go into the clinical intervention.

IASODs are also important because they highlight the difference between cognitive aids based on cognitive productivity tools versus task guidance (TG). Apps and other productivity tools make the individual active in applying a prosthetic intervention tool to a specific activity. The tool can provide the environment for an individual to demonstrate for the first time that s/he has the unexpected ability to perform a subtask, and giving the therapist and designer an opportunity to recognize the IASOD. In contrast, the TG approach applies a model that treats the IASOD as an error condition, and acts to "correct" the individual's behavior, forcing the individual to use the TG series of steps even if the individual would organize the activity differently. Not only does the TG approach induce the individual passive, but when there are opportunities for the individual to be active, the TG approach has already incorporated those subtasks into its plan.

9.3.4   ISLANDS OF DEFICITS IN SEAS OF ABILITIES (IDSOAS)

A problem found with cognitive rehabilitation provided in the clinic is that many of the cognitive advances made the clinic are not translated into advances in the patient's settings. IDSOAs provide an explanation. Functional rehab in the clinic uses simulations of an activity, rather than actual activity itself. These simulations are often carried out in an ADL idealized suite, and as a result, many barriers found in the natural environment have been removed. And the simulation clinic-based instance of the activity doesn't include critical subtasks—and cognitive dimensions—that the individual must perform in the natural environment. The missing subtasks include IDSOAs that have yet to be identified by the patient's clinicians.

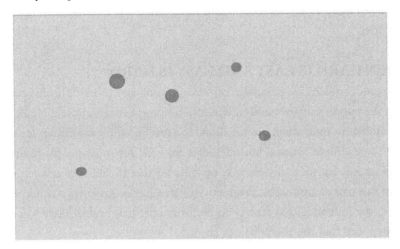

Figure 9.2: Islands of deficits in seas of abilities (ADSOA).

Both of the islands anomalies—IDSOAs and IASODs—are reasons why cognitive rehabilitation in the natural environment is superior to cognitive rehabilitation in the clinic. Additionally, productivity software operates at a sufficiently fine granularity that both islands will have an opportunity to appear, and be recognized by clinicians and designers. Both islands may also be a manifestation of function-specific brain-networks.

9.3.5   PATIENT CREATIVITY IN DEVELOPING FUNCTIONALITY IN SOFTWARE

Patient creativity in developing functionality in software is one of the substantial abilities unanticipated in TBI patients (an anomaly) described in the beginning of this section. Patient creativity is important in the process of cognitive rehabilitation, of increasing the way cognitive skills are combined in dealing with everyday activities. The structure of productivity software is well suited to an individual's discovering novel ways of combining features into new patterns (cf. Desouza et al., 2007; Schmitz et al., 2010).

In the context of cognitive disabilities, "Patient creativity in developing functionality in software" can refer to either of two scenarios. The first involves using a combination of application features in a way that hasn't been taught or trained. It is something that is novel to this individual user/patient. The second scenario is one where a novel use of features isn't included in the application's documentation, including templates and examples. Gail's proofreading of merchant data, by printing checks for $0 is an excellent example of this because the application designer hadn't anticipated that use of features. It could also be that the designers viewed that functionality as "illegal," and placed error traps to enforce their restrictions on functionality. For the context of cognitive disabilities, the key point is the innovative and novel use of application features to support one of the user's activities.

9.3.6 WHEN HARD IS EASY AND EASY IS HARD

Rehabilitation uses the staircase paradigm of moving from easy to hard. In this paradigm, the individual is given the opportunity of building up skills on progressively more difficult tasks. This is also seen as a way of determining when an individual has reached their maximum level of performance.

But for some people, what is considered to be hard for most people is easy for them. Additionally, what is easy for most people may be hard for them. There are several important clinical implications of this pair of anomalies. Productivity software use provides a window into "when hard is easy" because the individual has leeway to perform activities, and is likely to perform a "hard is easy" activity in a way that the therapist can see it.

To support Hugh's reading comprehension, we decided to obtain the bank's weekly loan review, a terse compilation for analysts and managers. This was far easier for Hugh than reading *The Wall Street Journal*. On the other hand, when asked to pick up a quart of milk and a loaf of white bread at a convenience store, he was overwhelmed, and left the store without anything. Here he demonstrated an example of "what is easy is hard," especially in comparison to his abilities with the weekly reports.

Laura would ordinarily have fallen through the cracks. She hadn't completed high school and was limited at a number of homemaking tasks. However she has a strong work ethic, and was able to return to work. Indeed, on returning to work nine months after her TBI, she was able to resume as if she had never left.

9.3.7 "OUT-OF-SEQUENCE" REHABILITATION

Cognitive rehabilitation has used the "staircase" approach. "Hard is easy and easy is hard" is an example of when—paradoxically—that approach fails. In the worst case, the therapist is unaware that the patient can do "hard" but fails at "easy," and the patient is mislabeled as low functioning because the "hard" behavior isn't observed within the testing or in-clinic therapy.

Patient-centered design will allow the patient to show their strengths, and develop goals that are appropriate for the higher level of functioning.

Return to work is typically the last step in TBI rehab, often involving vocational rehabilitation.

We have found it advantageous to have the patient return to the workplace early, and to light duty if organization procedures permit. The patient may only return for an hour or two a day, a varying number of days a week. A return to the workplace has an important social component for the patient. The individual is away from the clinical setting, and in a more normal social environment. Also, work activities are likely to be over-learned, with a good deal of routine; daily cognitive challenges at work not the norm. The patient visiting is more likely to be able to follow workplace chatter, and that can be very reassuring to the individual experiencing cognitive disabilities.

9.3.8   OTHER INSTANCES OF ANOMALIES IN BEHAVIOR

There are a number of anomalies that do not fit neatly into categories and are illustrated in various case studies below.

Roy was able to rapidly increase his time horizon. When he was accepted as a research subject, he was only able to work with time frames up to two weeks away. After using his initial intervention for a month, his time horizon had expanded to two months, while continuing to work with his therapist of three years.

Hugh, when evaluated at seven-weeks post injury, was in a cognitive fog for most contexts. However, when the topic changed to his occupation, his body language changed, his ability to converse changed, and the details and anecdotes changed. He seemed to be functioning at two cognitive levels.

Laura's recovery was remarkable, and presented a set of anomalies. She experience rapid cognitive and physical recovery from severe and profound disabilities. Hers was the most dramatic recovery. However, a number of other patients in mini case studies below experienced dramatic and unexpected degrees of recovery, including Gail, Suedell, Hugh, and Beth.

Dick's recovery was also remarkable. A stroke produced profound episodal memory failure, leaving him unable to perform activities on a daily schedule, and 15 months of cognitive rehabilitation was unable to address it. He recovered the ability to perform, drive himself around the community, and do volunteer work in the community and projects at home. Another anomaly involved his working memory, before his episodal memory became normal. He was able to work for two hours on a technical actuarial problem and his working memory supported the activity. When he finished the session, his memory of the session disappeared. A "Back to the Future" task strategy—leaving notes for picking up the task later—was developed so that he could resume where he had finished.

Gail had severe and profound cognitive deficits from a relatively mild automobile accident. After she had been using her prosthetic software for about six months, she mentioned that she had begun recognizing her face in the mirror when she was putting on lipstick. Prior to that, her lips were merely a target for her and she didn't experience them as hers.

Proficiency at Computer Use but Poor Reading Abilities

Zack, a high school sophomore with reading and other learning disabilities, was assigned to the Resource Room, the lowest classroom educational track. His reading tested at a fifth grade level, yet he had a substantial proficiency with computers. He provided tech support for his therapist during sessions. The intervention had several components. He was able to substantially increase his reading comprehension and writing skills in two-and-a-half months. An informal survey revealed a pattern with some students with major reading deficits: an ability to learn computer skills coupled with poor reading.

9.4   OPPORTUNITIES

Cognitive assistive technologies, when used in a modality of cognitive rehabilitation therapy, have many opportunities for research and professional practice.

9.4.1   EXPLORING PATIENT-CENTERED DESIGN

The CAT therapy area has rich potential for HCI researchers with limited budgets. This is an area where talent gets rewarded even when resources are minimal. A case in point is the Institute's 1992 study of three subjects—Suedell, Roy, and Sarah (Chapter 4). Over the course of two months, two researchers trained two therapists to do therapy with a radically new technology. Three of their patients were able to achieve remarkable results quickly on disabilities that had failed to respond to extended rehab efforts. This was accomplished with antique development tools and computing environments. The results were achieved so rapidly that the study scale was expended and a total of eight therapy goals were achieved for the three-month period. Today's software development environment requires so much less effort. Substantial and interesting results are likely to be attainable.

Patient-Centered Design provides a model that is adaptable to different kinds of cognitive disabilities; see Chapter 1 for a typology of cognitive disabilities. Patient-Centered Design dovetails with Cognitive Prosthetics Telerehabilitation.

PCD is efficient for the HCI designer in terms of both functionality and UI. The designer only has to focus on the needs of the individual patient/user, and only for one priority therapy goal at a time. Functionality can be designed into an application piecemeal, and there is the likelihood that functionality will be useful for other patients. Similarly, the UI, the most sensitive design element, only needs to serve that patient, who is in the best position to inform UI design. Working with several patients at a time is tractable.

The therapist is a key collaborator. S/he adds value to the use of CAT, and the HCI's researcher adds value to cognitive rehabilitation therapy. We chose to work with clinicians who were treating patients. These clinicians had patients that they had been treating for some time, and believed the patients had more potential than their current tools could achieve. We provided an opportunity for the therapists to take their patients to greater levels of functioning.

In working together, both therapist and designer will become aware of the differences in granularity that the two disciplines have in addressing the needs of the patient. Combining skill sets and models is what helps produce clinical and technical results.

9.4.2   METHODOLOGICAL ISSUES

The results from CPT-like therapy for TBI cognitive rehabilitation are promising, sufficiently promising to warrant study and rigor. There are a number of issues raised by the issue of a move toward rigor. The highest class of study of medical efficacy is the randomized control study (RTC) (SIGN 2008). However, Gillespie et. al. (2012), reviewing CAT in the *Journal of the International Neuropsychological Society,* suggests the single subject case study (SSCS) design has many benefits for advancing knowledge for CAT.

A Sufficient Number of Trained Therapists

A rigorous study requires a number of therapists who are fully competent to provide the CPT modality of therapy to patients. This basically means that an organization would need to train therapists to use this modality of therapy before undertaking or sponsoring a high-level efficacy study. Given the state of efficacy for present modalities of TBI rehabilitation, training therapists for CPT should not raise ethical barriers. SSCS designs could be used during the multi-level therapist training process.

Single Subject Case Studies

SSCSs have a long history in neurology and brain injury rehabilitation, and were acceptable high-level journals in those fields into the 1990s. Gillespie et al. suggest that SSCSs can be designed to have a higher level of evidence quality, and are another route to determining efficacy. To be at that higher level, SSCSs will need to have multiple data points at baseline, as well as multiple data points at intervention. They also propose either using standardized outcome measures or measures that have high inter-rater reliability and raters who are blind to the experimental hypotheses. They also call for "clear contextual data about the extent of support required" for the CAT. Efficacy is achieved by replicated studies.

9.4.3   MEASUREMENT FOR CPT

The Institute has used activity-outcomes as a means of assessing patient progress. This approach was able to match the scope of an intervention with the intervention outcome. Standardized clinical tests in the various clinical disciplines all seemed to have the same problem: the size of the phenomenon being measured was substantially larger than the size of the activity being addressed, e.g., successfully writing a letter (Leslie) was evaluated in terms of broad communications with broad others. This would result in the instrument's underestimating the patient outcome, particularly in the context of that individual's life.

Collaborators with clinical test-design capabilities may well be able to develop instruments that would not present bias by underestimating the impact of single and multiple CPT interventions.

In the second decade of the 21st century, neurophysiological measures including neuroimaging should be considered for studies. There are now functional neuroimaging technologies that are portable and capable of being used on many patient activities. One would hope that the progress that CPT patients have demonstrated would be of interest to some researchers.

Finally, one would hope that theories relating to HCI would be able to provide measures of patient performance changes.

9.4.4 IMPLICATIONS FOR ROBOTICS AND THE ACTIVE AND ENGAGED USER

A major stream of cognitive assistive technology research focuses on robotics for cognitive support of the individual with cognitive and physical disabilities. Most models have the robot guiding the user in performing an activity: the user plays a passive role in both robot design and activity execution. Our research suggests that the individual user is likely to be able to make improvements in his/her robot, particularly in the UI. In cognitive rehabilitation therapy, the patient needs to be active in order to make gains. In some rehabilitation, the robotic technology takes over, and the individual is not encouraged to become active or engaged.

The concept of augmenting cognition in the cognitively disabled individual through the use of a robot raises several possibilities. Robotic applications could be ready to provide assistance to the user in the taking of orders, providing information, and providing confirmation. In the area of robotics used to aid the cognitively disabled, it is likely that many of our results with computer-supported CPT would apply. For starters, (1) the activity supported should be important to the patient, (2) the UI is likely to present barriers to usability at first, and (3) the individual is likely to be able to provide valuable redesign ideas for various aspects of the UI. Moreover, the phrasing of the robotic dialogue may well be important for its implications about the user's competence. Our studies have shown instances in which what appears to be a user error in fact represents a significant step forward by the user. So it would be important for the robot not to correct the patient without asking a lot of questions first, unless delay would result in a life-threatening situation.

As we have seen, an important role of CAT is requiring patients to be active in their rehabilitation, which includes being active between therapy sessions. CAT must also try to engage the user emotionally in each activity that the technology supports. Engagement and activity repetition are important in inducing brain plasticity. The prospect of involving an interactive but subservient robot in this effort is intriguing, and given the amount of work that has already been done to develop social robots, the future looks promising.

9.4.5 ERRORLESS LEARNING STRATEGIES, COGNITIVE PROSTHETICS, AND EFFICIENT DELIVERY OF COGNITIVE SUPPORT SERVICES

Using the functional rehabilitation concept of everyday activities, errorless learning (Wilson et al., 1994) is an approach to training individuals with cognitive disabilities in how to perform an activity or use a device by reinforcing positive actions so that negative actions are substantially diminished. This approach has been widely adopted in cognitive rehabilitation therapy. Building on Wilson's work are two approaches of note involving CAT.

Alex Mihailidis (2011) has developed an approach he calls zero-effort technologies (ZETs). Many of Milhailidis's design principles are shared by the Institute. ZETs support functional activities in the real world, and their design allows minimal or no learning by the user, emphasizes UI design, encourages user involvement and participation, and requires personalization, which then complements the user's abilities. Milhailidis's results are particularly noteworthy because many of his studies have involved patients with severe dementia who have achieved impressive positive outcomes. The techniques used with patients easily fit within the bounds of therapy. Importantly, ZETs can be used with other devices and technologies. One could imagine that there would be many situations in which ZET and cognitive prosthetic telerehabilitation would provide better solutions than either could alone.

The other approach that builds on Wilson's work involves what is called systematic instruction (SI), which is also within the errorless learning model (Sohlberg and Turkstra, 2011). SI in this context is a training approach, personalized for each individual, that relies on explicit models, minimization of errors, the promotion of learner engagement, and carefully guided practice. Ehlhardt-Powell et al. (2012) applied SI to training in the use of a handheld computer by individuals with moderate to severe cognitive deficits from brain injury. In a study, each patient received 12 45-minute individualized instructional sessions aimed at supporting activities relevant to the patient. SI sessions in general involve substantial preparation by the trainer and therapist. These instructional sessions fall within the bounds of cognitive rehabilitation therapy. Techniques used in SI could likely contribute to cognitive prosthetics telerehabilitation therapy. It is also likely that using patient-centered design to personalize the patient's handheld computer could reduce SI training time.

9.4.6 NEW OPPORTUNITIES FOR CAT PRESENTED BY ADVANCES IN COGNITIVE NEUROSCIENCE

Cognitive neuroscience involves the senses and perception, plasticity, language, emotion and the social brain, higher cognitive functions, memory, attention, consciousness, and evolutionary biology. And for the study of cognitive disabilities, it would appear to be an important field to monitor, as it seems to supplement some of the clinical literature. This is a very broad and rapidly expanding scientific field. Graduate-level courses often use papers in lieu of books, often out of date on publication for research front workers. What is said in this book may also be somewhat out of date. With that caveat, Wilde et al. (2012) provide a primer of neuroimaging for workers in brain injury rehabilitation outcomes research (see also Molfese et al., 2011). Slobounov et al. (2012) address various kinds of neuroimaging for studying mild brain injury and concussion. Concussion has become an important vehicle for neuroimaging, and post concussion syndrome—concussions that don't fully resolve in ten days to a month—show many of the cognitive disabilities that have been traditionally associated with more severe brain injuries. Portable functional imaging is being used now in HCI usability studies (Solovey et al., 2012; Solovey et al., 2011). CAT specifically, and HCI generally, can take advantage of some of the emerging findings.

A high-visibility area of cognitive neuroscience is brain plasticity. It has broadly replaced, in a scientific context, the traditional model of the hard-wired brain. Brain plasticity is the ability of the brain to "rewire" itself in response to its environment, restoring a physical or cognitive function that had been damaged. Brain plasticity can be induced years post injury. This creates the potential for addressing a wide range of damaged functions. Three elements are thought to work together to induce brain plasticity: the importance of an activity to an individual, the individual's emotional engagement with that activity, and repetition of the activity. The Institute's cognitive prosthetics telerehabilitation focuses on a patient priority activity, encourages the user to become active and then engaged, and relies on repetition in the use of the functionality. Some case studies in this book have shown remarkable recovery of function years post injury and a relatively short time after cognitive prosthetic software had been introduced to therapy. At the time of these case studies, patient outcomes were considered anomalies. They were not explained by then-current theory, but the patient work products from their cognitive prosthetic software demonstrated a significant change in patient functioning and underlying abilities (e.g., Gail, Laura, Roy, Suedell, and Beth). This is strongly suggestive of brain plasticity but requires methodologically rigorous research to confirm.

9.4.7 DEVELOPING CLINICAL SERVICES

In today's dynamic medical environment, Cognitive Prosthetics Telerehabilitation offers both financial advantages and clinical advantages for traumatic brain injury and acquired brain injury cognitive rehabilitation. Cognitive rehabilitation for both TBI and ABI remain areas of poor outcomes if cognitive deficits do not resolve rather quickly. The Cognitive Prosthetics Telerehabilitation modality has demonstrated substantial results from a small patient population that had plateaued in other programs. Thus to a substantial extent, the "heavy lifting" on providing a commercial clinical service has been accomplished, reducing the risk for others.

For financial issues, CPT offers several advantages. Costs of providing cognitive rehab services are around 30% less than in-clinic rehab. The Institute's experience with third-party payers was a willingness to reimburse CPT therapy at market rates and for bundled services. Payers had no concerns about evidence-based studies, providing services via distance therapy regardless of contract language, or therapy across state lines. In terms of marketing, CPT lets the provider serve patients nearby or thousands of miles away: the catchment area becomes non-geographic.

There are additional costs for user support staff, CPT software, and videoconferencing software. These costs are modest, especially when allocated across patients, and a small fraction of the 30% gain from reduced costs.

With CPT, many procedures remain the same. Therapists come to the clinic to treat their patients. They will continue to use most of their clinical skills. The treatment plan structure and content remain fundamentally the same.

The therapy of course is different. Experienced clinicians can learn how to conduct CPT therapy sessions with two days of training. There will be two major adjustments. First, therapists will need to learn how to treat patients' actual activities, rather than simulations. Secondly, they will need to learn how to treat patients using cognitive prosthetic software in a shared online workspace. Therapists need on-site technical support by staff who are capable of working with clinicians. Therapists also need startup training. It is much more important that therapists have very good therapy skills than having talents in computing skills.

Exploring a new service involves the design of a pilot project. The case studies here suggest that the elements of Cognitive Prosthetics Telerehabilitation are synergistic, and outcomes derive from the interplay of the elements in Figure 6.1. There are several ways a limited-scale pilot can be designed. Gillespie et al. (2012) have suggested the importance of single subject case studies.

9.5   SUMMARY

This chapter has discussed conclusions, factors influencing outcomes, anomalies, and research opportunities. Since the 1980s, the author and the staff of the Institute for Cognitive Prosthetics have focused on the cognitive disabilities domain, developing technologies and techniques for helping individuals recover from brain injury. HCI models and methodologies were well suited to exploring the domain. CAT software was adopted by clinicians, who have been able to use it as a therapy tool to achieve cognitive gains in brain injury patients whose recovery had plateaued—in other words, a partial cure. Patient-Centered Design arose from the need to consider clinical factors in designing CAT for each patient. Cognitive Prosthetics Telerehabilitation evolved as a set of therapist tools, and as software tools for use during the cognitive recovery process. The user interface emerged as the most important area of software design for individuals with TBI and other brain injuries, and the patient emerged as the one best able to redesign and refine the UI so that it is effective and intuitive to the user. The ability to participate in designing their UIs is also an expression of the abilities of individuals with cognitive disabilities, even if the disabilities are profound and severe. Neuroscience and computing are both rapidly expanding and together hold great promise for treating cognitive disabilities. Research opportunities have been suggested for HCI researchers and for clinical services. Work on cognitive disabilities provides both emotional rewards and a fascinating view of the world's most complex information processing environment, the brain.

Bibliography

Abbot, Nan; Foschini, Maria; Cole, Elliot. (1989) Word Processing as a Compensatory Device in the Traumatically Head-Injured Survivor. In *Cognitive Rehabilitation*, January/February, pp . 36-38. 2

Alm, N., Astell, A., Ellis, M., Gowans, G., & Campbell, J. (2008). 2004). A cognitive prosthesis and communication support for people with dementia. *Journal of Neuropsychological Rehabilitation*, 14(1&2), 117-134. DOI: 10.1080/09602010343000147

Alm, N., Astell, A., Gowans, G., Ellis, M., Dye, R., Vaughan, P., & Riley, P. (2011.). Cognitive prostheses: Findings from attempts to model some aspects of cognition. Universal Access in Human-Computer Interaction. Design for All and Inclusion - 6th International Conference, UAHCI 2011, 275–284. 7, 8

Alm, N., Dye, R., Gowans, G., Campbell, J., Astell, A., and Ellis, A. (2007). A communication support system for older people with dementia. *IEEE Computer, , 40(5)*, 35-41. DOI: http://dx.doi.org/10.1109/MC.2007.153.

American Stroke Association. Retrieved from `http://www.strokeassociation.org/STROKEORG/AboutStroke/Impact-of-Stroke-Stroke-statistics_UCM_310728_Article.jsp.` 4

Arnott, J., Alm, N., & Waller, A. (1999). Cognitive prostheses: Communication, rehabilitation and beyond. IEEE SMC'99 Conference Proceedings. *IEEE International Conference on Systems, Man, and Cybernetics*, 6(October), 346–351. DOI:10.1109/ICSMC.1999.816576. 7

Astell, A. J., Ellis, M. P., Bernardi, L., Alm, N., Dye, R., Gowans, G., & Campbell, J. (2010). Using a touch screen computer to support relationships between people with dementia and caregivers. *Interacting with Computers*, 22(4), 267–275. DOI:10.1016/j.intcom.2010.03.003. 7, 89

Baecker, R. (2007). A taxonomy of technology for cognition. In *Proceedings of the 2nd International Conference on Technology and Aging*. PMid:18249055.

Baecker, R. (2008). Designing technology to aid cognition. In *Proceedings of the 10th international ACM SIGACCESS conference on Computers and accessibility* (Assets '08), 1-2. New York, NY: ACM. DOI: 10.1145/1414471.1414473. 112

Bellin, E., Dubler, N.N. (2001). The quality improvement-research divide and the need for external oversight. *American Journal of Public Health,* 91(9):1512-15.

Bigler, E. and Maxwell, W. (2012) Neuropathology of mild traumatic brain injury: relationship to neuroimaging findings." *Brain Imaging and Behavior*: 1-29. DOI: 10.1007/s11682-011-9145-0

Bigler, E. (1988) *Diagnostic Clinical Neuropsvchology*, Revised Edition, Austin, TX: University of Texas Press. PMid:21901424

Brain Injury Association. (n.d.) USA epidemiological data. Retrieved from `http://www.bi-ausa.org/about-brain-injury.htm`. 4

Brooks, Neil. (1984) *Closed Head Injury*. Oxford: Oxford University Press, 1984. 9

Burnham, A., Meredith, L., Helmus, T., Burns, R., Cox, R., D'Amico, E., Martin, L., Vaiana, M., Williams, K., and Yochelson, M. (2009) Systems of care: Challenges and opportunities to improve access to high quality care. In Tanielian and Jaycox (Eds.). *Invisible Wounds of War*. Santa Monica: RAND. 4

Card, S. K., Moran, T. P., and Newell, A. (1983). *The Psychology of Human-Computer Interaction*. Hillsdale, NJ: Lawrence Erlbaum Associates. 43, 110

Carmien, S. P. (2010). Socio-technical environments and assistive technology abandonment. In *Socio-technical Networks: Science and Engineering Design*. Taylor & Francis LLC, CRC Press. DOI: http://dx.doi.org/10.1201/b10327-8. 89

Carmien, S. P., Augustin, S., & Fischer, G. (2008). Design, adoption, and assessment of a socio-technical environment supporting independence for persons with cognitive disabilities. Beyond end-user programming. *Proceedings of ACM CHI 2008 Conference on Human Factors in Computing Systems, 1, 597-606*. DOI: 10.1145/1357054.1357151. 24, 42

Carmien, S.P. and Fischer, G. (2005). Tools for living and tools for learning. In *11th International Conference on Human-Computer Interaction (HCII 05)*. Published on CD-ROM by Lawrence Erlbaum Associates, Inc. ISBN: 0-8058-5807-5. 76

Carroll, J. and Carrithers, C. (1984). Training wheels in a user interface. *Communications of the ACM, 27*(8), 800-806. DOI:10.1145/358198.358218 `http://dx.doi.org/10.1145/358198.358218`. 57

Centers for Disease Control and Prevention. (n.d.). *Traumatic brain injury reports and fact sheets*. Retrieved from `http://www.cdc.gov/traumaticbraininjury/factsheets_reports.html`

Centers for Disease Control and Prevention. (n.d.) *Autism spectrum disorders: data and statistics*. Retrieved from `www.cdc.gov/ncbddd/autism/data.html`. 4

Cole, E. (1999) Cognitive Prosthetics: an overview to a method of treatment. *NeuroRehabilitation 12*(1): 39 – 51. 7, 78, 112

Cole, E., and Dehdashti, P. (1998). Computer-based cognitive prosthetics: Assistive technology for the treatment of cognitive disabilities. In *Assets 98: The Third International ACM Conference on Assistive Technologies*. New York, NY: ACM Press. DOI: 10.1145/274497.274502

Cole, E., and Dehdashti, P. (1990) Interface Design as a Prosthesis for Individuals with Brain Injuries. *SIG/CHI Bulletin*, 22(1): 28-32. 3, 28

Cole, E., Dehdashti, P., and Yonker, V. (1997). Treatment of plateaued traumatic brain injury patients by using computer-based cognitive prosthetics: A field study. NIH Neural Prosthesis Workshop, October 15-17, 1997. Available at `http://www.brain-rehab.com/pdf2/Cole_NIH_97_Neural_Prosthesis_Workshop_abstract.pdf`. 55

Cole, E., Dehdashti, P., Petti, L., Angert, M. (1994a). Participatory design for sensitive interface parameters: Contributions of traumatic brain injury patients to their prosthetic software. In *CHI '94 Conference Companion on Human Factors in Computing Systems*, 115-116. DOI: 10.1145/259963.260092. 47

Cole, E., Dehdashti, P., Petti, L., Angert, M. (1994b). Design and outcomes of computer-based cognitive prosthetics for brain injury: A field study of 3 subjects. *Neurorehabilitation*, **4**(3). 46, 47

Cole, E., Dehdashti, P., Matthews, M., and Petti, L. (1994c). Rapid functional improvement and generalization in a young stroke patient following computer-based cognitive prosthetic intervention. Presented at NIH Neural Prosthesis Workshop, Bethesda, MD, October 19-21, 1994.

Cole, E. and Dehdashti, P. (1992) Prosthetic software for individuals with mild traumatic brain injury: A case study of client and therapists. In *Proceedings of the RESNA International '92 Conference*, 170-172. PMid:20561240.

Cole, E., Dehdashti, P., Petti, L. (1993) Implementing complex compensatory strategies for brain injury. Presentation to the American Congress of Rehabilitation Medicine 70th Annual Meeting, Denver, CO, June 25-27, 1993. Abstract published in *Archives of Physical Medicine and Rehabilitation*, 74(6), 672.

Cole, E., Ziegmann, M., Wu, Y., Yonker, V., Gustafson, C., Cirwithen, S.. (2000). Use of "therapist friendly" tools in cognitive assistive technology and telerehabilitation. *RESNA Annual Conference Proceedings*, 31–33. 105

Cole, E., Bergman, M., and Dehdashti, P. (1988). Increasing personal productivity of adults with brain injuries through interface design. *ACM SIGCHI Bulletin*, 20(2), 32. DOI: 10.1145/54386.1046550 `http://dx.doi.org/10.1145/54386.1046550`. 2, 3

Constantinescu, Gabriella A ; Theodoros, Deborah G ; Russell, Trevor G ; Ward, Elizabeth C; Wilson, Elizabeth C ;and Wootton Richard. (2010. Home-based speech treatment for Parkinson's disease delivered remotely: a case report. *J Telemed Telecare* March 2010 16:100-104; 113

Cramer, Steven C; Mriganka Sur, Bruce H. Dobkin, Charles O'Brien, Terence D. Sanger, John Q. Trojanowski, Judith M. Rumsey, Ramona Hicks, Judy Cameron, Daofen Chen, Wen G. Chen, Leonardo G. Cohen, Christopher deCharms, Charles J. Duffy, Guinevere F. Eden, Eberhard E. Fetz, Rosemarie Filart, Michelle Freund, Steven J. Grant, Suzanne Haber, Peter W. Kalivas, Bryan Kolb, Arthur F. Kramer, Minda Lynch, Helen S. Mayberg, Patrick S. McQuillen, Ralph Nitkin, Alvaro Pascual-Leone, Patricia Reuter-Lorenz, Nicholas Schiff, Anu Sharma, Lana Shekim, Michael Stryker, Edith V. Sullivan and Sophia Vinogradov. (2011) Harnessing neuroplasticity for clinical applications. *Brain* 134 (6): 1591-1609. doi: 10.1093/brain/awr039. 95

Danzl, M., Etter, N.,. Andreatta, R., Kitzman. (2012) Facilitating Neurorehabilitation Through Principles of Engagement. *Journal of Allied Health*, **41**(1), pp. 35-41(7). 89

Dawe, Melissa. (2006). Desperately seeking simplicity: how young adults with cognitive disabilities and their families adopt assistive technologies. In *Proceedings of the SIGCHI Conference on Human Factors in Computing Systems.* DOI=10.1145/1124772.1124943. 89

Dehdashti, P., and Cole, E. (1994) Interface design techniques for cognitive prosthetic software for individuals with traumatic brain injury. Presented at the 5th International Conference on Human-Computer Interaction. 47

Desouza, K. C., Awazu, Y., and Ramaprasad, A. (2007). Modifications and innovations to technology artifacts. *Technovation*, **27**(4), 204–220. DOI:10.1016/j.technovation.2006.09.002. 119

Ehlhardt-Powell, L., Glang, A., Ettel, D., Todis, B., Sohlberg, M. and Albin, R. (2012). Systematic instruction for individuals with acquired brain injury: Results of a randomised controlled trial. *Neuropsychological Rehabilitation*, **22**(1), 85–112. DOI: 10.1080/09602011.2011.640466. 125

Engelbart, D. (1962, October). *Augmenting human intellect: A conceptual framework.* SRI Summary Report AFOSR-3223, Prepared for: Director of Information Sciences, Air Force Office of Scientific Research. SRI International, hosted by The Doug Engelbart Institute. 28

Gajos, Krzysztof; Hurst, Amy; Findlater, Leah. (2012). Personalized dynamic accessibility. *Interactions* 19(2): 69-73. 6

Gazzaniga, M., Doran, K., and Funk, C. (2009)). Looking toward the future: Perspectives on examining the architecture and function of the human brain as a complex system, In M. Gazzaniga, M. (Ed.), *The Cognitive Neurosciences,* 4th ed. (1242–1247). Cambridge: MIT Press. 23

Gazzaniga, Michael, Ed. (2009) *The Cognitive Neurosciences*, 4th ed. Cambridge, MIT Press.

Gillespie, A., Best, C., and O'Neill, B. (2012). Cognitive function and assistive technology for cognition: a systematic review. *Journal of the International Neuropsychological Society : JINS*, **18**(1), 1–19. doi:10.1017/S1355617711001548. 7, 13, 123, 127

Gómez, J., Montoro, G., Haya, P. a., Alamán, X., Alves, S., and Martínez, M. (2012). Adaptive manuals as assistive technology to support and train people with acquired brain injury in their daily life activities. *Personal and Ubiquitous Computing*. DOI:10.1007/s00779-012-0560-z.

Gould, J. and Lewis, C. (1985) Design for usability: Key principles and what designers think. *Communications of the ACM*, **28**(3), 300–311. 32

Green, J.L. (2011). *The ultimate guide to assistive technology in special education: Resources for education, intervention, and rehabilitation*. Waco Tx: Prufrock Press. 78

Horn, L. J; and Zasler, N. D. (1996) *Medical Rehabilitation of Traumatic Brain Injury*. Philadelphia: Hanley & Belfus.

(IOM) Institute of Medicine. (2011). *Cognitive rehabilitation therapy for traumatic brain injury: Evaluating the evidence*. Washington, DC: The National Academies Press. Available from `http://www.nap.edu/openbook.php?record_id=13220`

Kane, Shaun K; Unam-Church, Barbara; Althoff, Kyle, and McCall, Denise (2012). What we talk about: designing a context-aware communication tool for people with aphasia. *Assets* 12 Proceedings, 39-45.

Kapur, N., Glisky, E. L., and Wilson, B. A. (2004). Technological memory aids for people with memory deficits. *Neuropsychological Rehabilitation*, **14**(1/2), 41 – 60. DOI:10.1080/09602010343000138 PMCid:1191901.

Kim, H. J., Burke, D. T., Dowds, M. M., Boone, K.A., and Park, G. J. (2000). Electronic memory aids for outpatient brain injury: Follow-up findings. *Brain injury : [BI]*, **14**(2), 187–96. Retrieved from `http://www.ncbi.nlm.nih.gov/pubmed/10695574`.

Kirsch, N. L. Levine, S.P., Fallon-Krueger, M., Jaros, L.A. (1987). The Microcomputer as an "orthotic" device for patients with cognitive deficits. *Journal of Head Trauma Rehabilitation*, **2**(4) pp. 77-86. `http://dx.doi.org/10.1097/00001199-198712000-00012`. 6

Kirsch, N. L., Shenton, M., Spirl, E., Rowan, J., Simpson, R., Schreckenghost, D., & LoPresti, E. F. (2004). Web-based assistive technology interventions for cognitive impairments after traumatic brain injury: A selective review and two case studies. *Rehabilitation Psychology*, **49**(3), 200–212. DOI:10.1037/0090-5550.49.3.200. 113

Langley, G.L., Nolan K.M., Nolan T.W., Norman C.L., and Provost L.P. (2009) *The Improvement Guide: A Practical Approach to Enhancing Organizational Performance* (2nd edition). San Francisco: Jossey-Bass Publishers. 74

Lequerica, A. H., and Kathleen K. Therapeutic engagement: A proposed model of engagement in medical rehabilitation. *American Journal of Physical Medicine & Rehabilitation*, 89(5), 415-422. DOI: 10.1097/PHM.0b013e3181d8ceb2 `http://dx.doi.org/10.1097/PHM.0b013e3181d8ceb2`. 89

Lewis, C. 2007. Simplicity in cognitive assistive technology: A framework and agenda for research. *Universal Access in the Information Society*, **5**(4), 351-361. DOI: 10.1007/s10209-006-0063-7.

Lewis, C. (1986). A model of mental model construction. In *Proceedings of the SIGCHI Conference*, 306–313. 45

Lezak, M. D., Howieson, D.B., Bigler, E.D., and Tranel, D. (2012). *Neuropsychological Assessment*, (5th ed.). New York: Oxford University Press. 8, 17

Licklider, J.C.R. (1960). Man-computer symbiosis. *IRE Transactions on Human Factors in Electronics*, 1, 4-11. DOI: 10.1109/THFE2.1960.4503259. 28

Linebarger, Marcia; McCall, Denise; Virata, Telana; and Berndt, Rita. (2007). Widening the temporal window: Processing support in the treatment of aphasic language production. *Brain and Language* 100(1): 53-66. 78

LoPresti, E. F., Bodine, C., and Lewis, C. (2008). Assistive technology for cognition [Understanding the needs of persons with disabilities] *IEEE Engineering in Medicine and Biology Magazine*, **27**(2), 29–39. DOI:10.1109/EMB.2007.907396. 7, 13

Lopresti, E., Mihailidis, A., and Kirsch, N. (2004). Assistive technology for cognitive rehabilitation: State of the art. *Neuropsychological Rehabilitation*, **14**(1-2), 5–39. DOI:10.1080/09602010343000101. 7, 13, 42

Mayer, NN., Keating, D., and Rapp, B. (1986). Skills, routines, and activity patterns of daily living: A functional nested approach. In B. Uzzell and Y. Gross (Eds.). *Clinical Neuropsychology of Intervention*. Boston: Martins Nijhoff. `http://dx.doi.org/10.1007/978-1-4613-2291-7_10` PMid:18836062 PMCid:2602740. 16

McCall, D., Telana, V., Marcia, C.L., and Berndt, R. (2009). Integrating technology and targeted treatment to improve narrative production in aphasia: A case study. *Aphasiology*, **23**(4), 438-461. DOI: 10.1080/02687030701818000. 78

McGrenere, J., Baecker, R.M., and Kellogg, B. (2007). A field evaluation of an adaptable two-interface design for feature-rich software. *ACM Transactions on Computer-Human Interaction (TOCHI)*, **14**(1). DOI: 10.1145/1229855.1229858.

McGrenere, J., Sullivan, J., and Baecker, R. (2006). Designing technology for people with cognitive impairments. *CHI Workshop, Extended Abstracts of ACM CHI 2006*, 1635-1638. DOI: 10.1145/1125451.1125750

McGrenere, J., Davies, R., Findlater, L., Graf, P., Klawe, M., Moffatt, K., Yang, S. (2003). Insights from the aphasia project: Designing technology for and with people who have aphasia. In *Proceedings of the 2003 ACM conference on Universal Usability*, 112–118. DOI: 10.1145/957205.957225 http://dx.doi.org/10.1145/957205.957225. 112

Merzenich, Michael M. Personal Communciation. October 14, 1993. 98, 116

Mihailidis, A., Boger, J., Hoey, J., and Jiancaro, T. (2011) *Zero effort technologies: Considerations, challenges, and use in health, wellness, and rehabilitation.* San Rafael, CA: Morgan & Claypool Publishers. 27, 112, 125

Mihailidis, A. (2008). The efficacy of an intelligent cognitive orthosis to facilitate handwashing by persons with moderate to severe dementia. *Neuropsychological Rehabilitation*, **14**(1-2), 135–171. DOI:10.1080/09602010343000156.

Mihailidis, A., Fernie, G. R., and Barbenel, J.C. (2001). The use of artificial intelligence in the design of an intelligent cognitive orthosis for people with dementia. *Assistive Technology: The Official Journal of RESNA*, 13(1), 23–39. DOI:10.1080/10400435.2001.10132031.

Moffatt, K., and Davies, R. (2004). The Aphasia project: Designing technology for and with individuals who have aphasia." *ACM SIGACCESS Accessibility and Computing*, **80**, 11–17. 112

Molfese, D., Molfese, V., and Garrod, K. (2012). Evidence of dynamic changes in brain processing from imaging techniques: Implications for interventions for developmental disabilities. In Molfese, D., Breznitz, Z., Berninger, V., and Rubinsten, O. (Eds.). *Listening To Many Voices: Reading, Writing, Mathematics and The Brain.* New York, NY: Springer. DOI: 10.1007/978-94-007-4086-0_2. 126

National Institute of Neurological Diseases and Stroke. (n.d.) Retrieved from www.ninds.nih.gov/dementias/detail_dementia.html

National Research Council, Institute of Medicine. 2011. *Cognitive Rehabilitation Therapy for Traumatic Brain Injury: Evaluating the Evidence.* Washington, DC: The National Academies Press. http://www.nap.edu/openbook.php?record_id=13220. 16

Newell, A.F., (2011). Design and the Digital Divide: Insights from 40 Years in Computer Support for Older and Disabled People. San Rafael, CA: Morgan & Claypool Publishers. 7, 61

Norman, Donald A. (2011) *Living with Complexity Cambridge.* MIT Press. 89

O'Neill, B., Moran, K., & Gillespie, A. (2010). Scaffolding rehabilitation behaviour using a voice-mediated assistive technology for cognition. *Neuropsychological Rehabilitation*, **20**(4), 509–27. DOI:10.1080/09602010903519652. 69

Ogrinc, G., Mooney, S.E., Estrada, C., Foster, T., Goldmann, D., Hall, L.W., Watts, B. (2008). The SQUIRE (Standards for QUality Improvement Reporting Excellence) guidelines for quality improvement reporting: Explanation and elaboration. *Qual Saf Health Care*, **17**(i13-i32). http://dx.doi.org/10.1136/qshc.2008.029058. 74

Pollack, M. E., Mccarthy, C. E., and Ramakrishnan, S. (2003). Execution-time plan management for a cognitive orthotic system. In M. Beetz et al. (Eds.). *Plan-Based Control of Robotic Agents* (pp. 1–14). New York, NY: Springer.

Raskin, Sarah S. (2011) *Neuralplasticity in Rehabilitation*, Guilford Press. 116

RESNA Instructional Courses. (2000). Computer applications: Applications for cognitive disabilities. http://www.resna.org/. 7

Robertson, I.H., McMillan, T.M., MacLeod, E., Edgeworth, J., and Brock, D. (2002). Rehabilitation by limb activation training reduces left-sided motor impairment in unilateral neglect patients: A single-blind randomised control trial. *Neuropsychological Rehabilitation* **12**: 439. DOI:10.1080/09602010244000228. 116

Rogers, Everett (2003) *Dffusion of Innovatrions*, 5th ed. New York: Free Press. PMid:22669496. 77

Sassaroli, A., Zheng, F., Hirshfield, L.M., Girouard, A., Solovey, E.T., Jacob, R.J.K., and Fantini, S. (2008). Discrimination of mental workload levels in human subjects with functional near-infrared spectroscopy. *Journal of Innovative Optical Health Sciences*, **1**(2) 227-237. DOI: 10.1142/S1793545808000224

Schell, T.L., and Marshall, G.N. (2009). Survey of individuals previously deployed for OEF/OIF. In Tanielian and Jaycox (Eds.). *Invisible Wounds of War*. Santa Monica: RAND. 4

Scherer, Marcia (2012). *Assistive Technologies and other supports for People with Brain Impairment*. New York: Springer. 13, 89

Schmitz, K., Webb, K., and Teng, J. (2010). Exploring technology and task adaptation among individual users of mobile technology exploring technology and task adaptation. In *Proceedings of ICIS*, 57-57. 119

Silver, J., McAllister, M., Thomas W, and Yudofsky, S.C. (2011). *Textbook of Traumatic Brain Injury* (2nd ed.). Arlington, VA: American Psychiatric Publishing. PMid:20723194. 15

SIGN Scottish Intercollegiate Guidelines Network (2008). *SIGN 50: A guideline developer's handbook*. Edenbutgh: Scotish Intercollegiate Guidelines Network. 123

Slobounov, S., Gay, M., Johnson, B., and Zhang, K. (2012). Concussion in athletics: Ongoing clinical and brain imaging research controversies. *Brain Imaging and Behavior*, **6**(2), 224-243. DOI: 10.1007/s11682-012-9167-2. 126

Sohlberg M. and Turkstra, L. (2011). *Optimizing Ccognitive Rehabilitation*. New York, NY: Guilford Press. 20, 27, 125

Solovey, E., Schermerhorn, P., Scheutz, M., Sassaroli, A., Fantini, S., and Jacob, R. (2012). Brainput: Enhancing interactive systems with streaming fnirs brain input. In *CHI '12 Proceedings of the SIGCHI Conference on Human Factors in Computing Systems*, 2193-2202. DOI: 10.1145/2207676.2208372. 126

Solovey, E., Lalooses, F., Chauncey, K., Weaver, D., Parasi, M., Scheutz, M., Jacob, R. (2011). Sensing cognitive multitasking for a brain-based adaptive user interface. In CHI '11 Proceedings of the SIGCHI Conference on Human Factors in Computing Systems, 383-392. DOI: 10.1145/1978942.1978997. `http://delivery.acm.org/10.1145/2210000/2208372/p2193-solovey.pdf`. 126

Stein, D., Brailowsky, S., and Bruno, W. (1995). *Brain Repair*. New York, NY: Oxford University Press.

Theodoros, D. (2011). Telepractice supported Delivery of LSCT®Loud. *Prospectives on Neurophysiology and Neurogenic Speech and Language Disorders* **21**, 107-119 October. DOI: 10.1044/nnsld21.3.107. 78, 79, 116

Tremaine, M., Sarcevic, A., Dezhi, W., Velez, M., Dorohonceanu, B., Krebs, A., Marsic, I. (2005). Size does matter in computer collaboration: Heterogeneous platform effects on human-human interaction. In *Proceedings of the 38th Annual Hawaii International Conference on System Sciences, HICSS '05*. DOI: 10.1109/HICSS.2005.543. 82

Walpaw, Jonathan R. (2012) Letter to the Editor: Harnessing neuroplasticity for clinical applications. *Brain* (2012) 135 (4): e215. doi: 10.1093/brain/aws017. 95

Wilde, E., Hunter, J., Bigler, E. (2012). A primer of neuroimaging analysis in neurorehabilitation outcome research" *NeuroRehabilitation*, **31**(3), 227–242. 126

Wilson, B. A. (1986). *Rehabilitation of Memory*. New York, NY: Guilford Press.

Wilson, B. A. (2000). Compensating for cognitive deficits following brain injury. *Neuropsychology Review*, **10**(4), 233–43.

Wilson, B. A. (2009). *Memory Rehabilitation: Integrating Theory and Practice*. Guilford Press.

Wilson, B. A., Baddeley, A. D., Evans, J. J., and Shiel, A. (1994). Errorless learning in the rehabilitation of memory impaired people. *Neuropsychological Rehabilitation*An International Journal, **4**(3), 307–326. DOI: 10.1080/09602019408401463. 125

Wilson, B.A., Emslie, H.C., Quirk, K., and Evans, J.J. (2001). Reducing everyday memory and planning problems by means of a paging system: A randomised control crossover study. *Journal of Neurology, Neurosurgery, and Psychiatry*, **70**: 477–482. doi:10.1136/jnnp.70.4.477. 116

Wobbrock, Jacob O. ; Kane, Shaun K.; Gajos, Krzysztof Z.; Harada, Susumu ; and Froehlich Jon. (2011). Ability-Based Design: Concept, Principles and Examples. *Transactions on Accessible Computing* (TACCESS), 3(3). DOI=10.1145/1952383.1952384. 42

World Health Organization, 2001. *International Classifications of Functioning, Disability, and Health (ICF)*. Available at `www.who.int/classifications/icf/en/`. 18

Wu, Michael. Personal Communication, August 18, 2012. 92

Wu, M. Baecker, R., and Richards, B. 2010. Field evaluation of a collaborative memory aid for persons with amnesia and their family members. *Proceedings of the 12th international ACM SIGACCESS*, 51-58. DOI=10.1145/1878803.1878815. 92, 112

Wu, M. Baecker, R., and Richards, B. (2004) Participatory design with individuals who have amnesia. In Proceedings of the eighth conference on Participatory design: Artful integration: interweaving media, materials and practices, PDC 04, 1, 214 – 223. DOI: 10.1145/1011870.1011895. 92, 112

Author Biography

Dr. Elliot Cole is the founder of the Institute for Cognitive Prosthetics. He brings his training in human-centered computing to developing technology and techniques that address cognitive disabilities from brain injury. The Institute's successful R&D efforts came from a multidisciplinary staff from clinical and computing specialties working closely together and focusing on the rehabilitation needs of the individual patient. This approach has generated deep knowledge of the cognitive disabilities computing domain. For over a decade, the Institute had a "lab" brain injury cognitive rehabilitation facility. Dr. Cole was an associate professor at Drexel University and a research associate at the University of Pennsylvania, where he is currently a Visiting Scholar.

Printed in the United States
by Baker & Taylor Publisher Services